Voices of Snipers

Eyewitness Accounts from the World Wars

John Walter

Greenhill Books

Voices of Snipers: Eyewitness Accounts from the World Wars

Greenhill Books

Greenhill Books
First published by Greenhill Books, 2022
Greenhill Books, c/o Pen & Sword Books Ltd,
47 Church Street, Barnsley, S. Yorkshire, S70 2AS
For more information on our books, please visit
www.greenhillbooks.com, email contact@greenhillbooks.com
or write to us at the above address.

CIP data records for this title are available from the British Library
ISBN 978-1-78438-627-6

Typeset by JCS Publishing Services Ltd
Typeset in 10.5/13pt Minion Pro
Back cover: Image credit Author's archives
Front cover: Image credit SA-kuva
Spine: Image credit Morphy Auctions
Printed and bound in Great Britain by
CPI Group (UK) Ltd, Croydon, CRO 4YY

Note on Russian transliteration

One of the problems of republishing Russian-language memoirs lies in Cyrillic transliteration. Many differing systems have been used over the years, some relying on character-by-character methods and others more on pronunciation. For example, Cyrillic 'E' is sometimes rendered as 'e' and sometimes as 'ye'.

Voices of Snipers uses a character-based system derived from British Standard 2979:1958, but the appended *Roza's Diary* retains the spellings of the original translation. This explains the occasional inconsistency in names such as Aleksandra Yekimova, 'Ekimova' on the standardised scheme.

CONTENTS

Foreword by Martin Pegler v

Preface ix

Our Chroniclers xi

Rise of the Sniper 1
- Snipers and Gallipoli 9
- Uneasy Peace, Certain War 12
- The Russian Story 13
- Sniping and Women 16
- Showdown in North Africa and the Mediterranean 20
- Up the Beach and into the Jungle 23
- Across Europe to the Reich 25

What Makes a Sniper? 28
- Body, Mind and Spirit 29

Recruiting the Sniper 38
- Allied Recruitment 48

Training the Sniper 54

Take your Weapons! 71
- Optical Sights 72
- Rifles for Snipers, 1914–18 76
- Rifles for Snipers, 1939–45 87

Trials and Tribulations 99
Personal-defence and Other Weapons 101
Ammunition 107
Accessories 114

Shoot If You Can! 119
Finding the Range 122
Into Position 125
Take Cover! 128
Hide and Seek 136
New Tricks 141
Graduation 143

Snipers in Combat 148
Into Action 152
Seeking the Target 160
Kill the Sniper 176
Tallies: Fact or Fiction? 186
In Their Honour … 190

After the Fight 196
Obscurity and Rehabilitation 198

Appendices 201
Appendix 1: A Typical Sniping Course 201
Appendix 2: Roza's Diary 204

Bibliography 251

FOREWORD

Virtually all of our learned history is second-hand. Historical events are always open to interpretation, even by those who were there at the time, and almost no one is ever able to present a cohesive account of what happened, or even why, until a considerable time after the event. This is never more so than when looking at military history. Understanding a campaign or a single battle can only be done in retrospect, when all of the facts, figures and accounts have been collated and sorted into a chronological history. The problem with this is that generally what the historian then presents is at best an overview of what happened based solely on the accounts of people who were seldom physically present. This is even more true in the twentieth century, when senior commanders rarely set foot on battlefields (Field-Marshal Haig never travelled closer than 16 miles to the Somme front during the battle) relying instead on intelligence reports, maps and photographs to plan and execute their strategies.

John Walter's *Voices of the Snipers* attempts to redress this by using the words and experiences of the snipers themselves, taken from a myriad of books, interviews and memoirs. To undertake such a project is no mean feat, for until well into the 1980s almost nothing was published by snipers on their invisible war. In most military circles it was regarded as a somewhat uncivilised way to wage war and although newspapers published accounts of sniping during and after the First World War, these were mostly embellished and seldom contained any direct quotes from the snipers themselves, who by choice were the most solitary soldiers on the battlefield and who preferred to remain publicly and privately invisible.

Their war was a strange contradiction, being both acutely personal whilst at the same time dispassionate. Unlike any other combatant, they looked at the man they were about to kill, for they literally held the power of life or death in their hands. Even that other solitary hero of the age, the fighter pilot, could convince himself he was shooting at mechanical objects, not human beings. Snipers never had the luxury of this mental shield, for they knew exactly what the results of their actions would be, and each had to learn to deal with it in their own way. Their war was as far removed from the lofty and indifferent accounts of the generals as it was possible to be. It was not until after the Vietnam War that any first-hand accounts from snipers began to emerge, acting as a balance to the grandiloquent histories of war. Even then, their vital part in defeating the enemy in a hidden war, in which there was no quarter, was largely ignored in terms of military awards.

Of the allied commanders who served in the Second World War, virtually every one received some form of military decoration acknowledging the part they played. Britain's highest scoring sniper, Harry Furness, who fought from the D-Day landings through to Germany and was twice wounded, received no more than the usual four service medals. He was fortunate that he even survived as casualty rates for allied snipers were over 50 per cent killed and wounded. He was the only original sniper in his battalion to reach Germany. For the Russians, their war was even harder; the bitter fighting at Stalingrad meant that a sniper's lifespan was expected to be only two weeks.

This was the gritty reality of the sniping war: figures that became lost in masses of statistics, but tell us nothing about the men who fought or the type of combat they experienced. *Voices* opens a Pandora's box of experiences and clears the 'fog of war' that made it impossible for any combatant to say coherently what had happened except for the few dozen yards on either side of them. They had no idea whether they were beating the enemy or not; if their own localised attack had proved successful, or if they were in fact about to be surrounded. The wider tactical implications were utterly lost on them. Every soldier fought a totally personal war, relying on their immediate comrades to protect their flanks, dealing with each threat as it appeared whilst trying to keep a clear head and a grip on their fear. How this affected them, at the time and in later years, is a fascinating story in itself.

There was another seldom talked about element to the sniping war: unlike ordinary infantrymen in combat, a sniper had little recourse to help if things went awry, for they often fought far in advance of their own lines and faced certain death if captured. It would not necessarily be quick either, particularly on the Eastern Front, and the carrying of a pistol was routine – a bullet in the head was far preferable to the alternatives. It is almost certainly true that very few returned to civilian life the same people that they were before the war, and a few touch on the problems that they faced post-war. Alcohol, depression and the now almost-universal post-traumatic stress disorder took their toll and it is impossible to know how many former snipers succumbed as a result. A few, such as Yuliya Zhukova, author of *Girl With A Sniper Rifle,* write openly about postwar problems. Others, such as Lyudmila Pavlichenco, 'Lady Death', battled alcohol and died prematurely. In a very moving end to his grim book, *Sniper On The Eastern Front*, Sepp Allerberger said that, despite surviving the war almost physically unhurt, his heart would remain hard for the rest of his life for 'the spectre of war would never let go of him.' This was true for thousands of former snipers, who mostly suffered in silence, unable to seek any professional help, and we would never know their remarkable and often terrifying stories were it not for authors such as John Walter, who must be congratulated on this impressive body of work. He has put together a cohesive history of the darkest and most secretive form of warfare in modern history, as told by those faceless, and often nameless, snipers who have always lived in the shadows.

Martin Pegler
2021

PREFACE

When – having worked on *Snipers at War*, the *Sniper Encyclopaedia* and Roza Shanina's diary – I was asked to put together an intimate view of the subject, I was uncertain if sufficient material existed to present one which was truly even-handed.

Availability of first-hand testimony clearly reflects the differing national views of snipers: from the almost total absence of Japanese recollections to the plethora of Russian memoirs from the Second World War. Comparatively little British and American material is readily available, probably owing to the once-common view of snipers as cold-blooded killers, and the many men who served the German forces are represented principally by the fascinating stories of only three leading marksmen: Josef 'Sepp' Allerberger, Matthäus Hetzenauer and Bruno Sutkus.

But can these recollections be trusted? Having read countless thousands of words, I feel that their veracity can sometimes be questioned. There are honourable exceptions, of course: Frederick Crum kept a day-by-day diary and the story of Bruno Sutkus, for example, is based on his surviving highly detailed wartime logbook.

Some of the memoirs are low-key personal testimonies, which for the most part ring true. Yet there are others in which the work of propagandists, political masters and exploitative editors can be discerned. Several Russian snipers subsequently pursued careers in which the state was directly involved, and their views can reflect an understandable tendency to avoid criticising the leadership of the USSR.

The atrocities that inevitably occur in wartime generally receive little more than passing mention in Russian accounts, but one of the German memoirs gives details with such precision that the material may have been deliberately sensationalised. While I can accept that the incidents occurred, claiming to remember what was said word-for-word, at second hand and often in a language foreign to the sniper, should be questioned.

Voices of Snipers elects to tell a story from beginning to end, from volunteering or impressment to death or discharge, illuminated by first-hand commentaries from not only snipers but also trainers such as Frederick Crum, Hesketh Prichard and Herbert McBride – shooting enthusiasts with skills proven in combat. The rise of sniping has been explained only briefly, much more detail being accessible elsewhere (see Bibliography). However, historical elements occasionally appear, to reflect the effects on sniping as the First World War stagnated in the trenches of the Western Front, or on the heights of Gallipoli peninsula, to the desperation of the Red Army's push across Europe and the island-by-island advance by Allied forces in the Pacific.

Inevitably, the selection of quotes, with so much material to choose from, has been essentially personal. But I hope that those that have been used will guide you, the reader, through the story in such a way that you feel you're actually partaking in the snipers' experience!

One of the most shocking elements of the story is the youth of many participants: still in their teens or early twenties, many men and women chose or were forced to endure the horrors of war. Some paid the ultimate price, while others made it to the end. For many, however, the psychological scars never healed, no matter how many reunions they attended and how much rehabilitation was undertaken.

My work could not have been completed without help. Consequently, apart from 'globally thanking' those who have allowed snippets of memoirs and personal testimonies to be used, I would like to express particular gratitude to Lionel and Michael Leventhal, David Foreman, Martin Pegler, Ian Skennerton, Charles Sasser, Leroy Thompson, Dina Nikolaeva, Yuliya Zhukova and Alla Begunova. And editor Steve Williamson found many things that had eluded the author!

For help with images: Hermann Historica, Morphy Auctions and Rock Island Auctions.

And, last but not least, all the usual suspects: Alison, Adam, Nicky, Findlay, Georgia and Holly, plus Amber and new addition Maggie!

John Walter

2021

OUR CHRONICLERS

Detailed biographical information concerning entries marked '*' will be found in *The Sniper Encylopaedia*.

Josef 'Sepp' Allerberger * (1924–2010), subject of Albrecht Wacker's *Sniper on the Eastern Front* (2005), joined the German Army in August 1943. Serving with Gebirgs-Jäger-Regiment 144, his tally of 257 kills from 1 September 1944 until the end of the war is second only to that of Matthäus Hetzenauer. Consequently, he was awarded the Knight's Cross of the Iron Cross on 20 April 1945. His memoirs were published in Germany in 2000 as *Im Auge des Jäger* but names were changed to preserve the anonymity of men who were still alive; however, the English-language edition correctly identified most of the participants.

Rem Altshuller, born in 1926 to a soldier and a school teacher, volunteered for service in Leningrad immediately after the German invasion of the USSR. Though under-age, he was able to persuade recruiters of the value of his apparent knowledge of German – Altshuller was Jewish, with Yiddish and German being sufficiently closely related to make his claim acceptable. He was also holder of the Voroshilov marksmanship badge, suiting him to the sniper's role for which he had soon qualified, but he spent much of his time as an interrogator as his units advanced to the Reich.

Nevill Alexander Drummond Armstrong (1874–1954), author of *Fieldcraft, Sniping and Intelligence* (1941), served as Chief Instructor of the 2nd Army School of Scouting, Observation and Sniping in 1915–16, Commandant of the Canadian Corps School of Scouting, Observation and Sniping (1917–18), and then as Chief Reconnaissance Officer. A member of the Royal Geographical Society, Armstrong was awarded the OBE for his services.

Charles Edwin Woodrow Bean (1879–1968), a war correspondent/historian, is now best remembered as editor of the 12-volume *Official History of Australia in the War of 1914–1918*, which includes material devoted to Gallipoli.

Pyotr Alekseevich Belyakov ✻, born in 1927, underwent military training during his last year at school. Conscripted in 1942, his prowess as a marksman ensured he was sent for training as a sniper. Beginning service near Stalingrad at the end of 1942, he reached the Dnepr river in November 1943 but was then wounded badly enough to cease sniping, with a tally said to have amounted to 147.

Alan Johnston Campbell (1895–1982), who gave details to John Hamilton for use in *Gallipoli Sniper*, served as a trooper with the Australian 2nd/5th Light Horse.

Daniel Webster Cass Jr (1916–2011), contributor to Charles Sasser and Craig Roberts's *One Shot – One Kill* (1990), served in the US Marine Corps as a sniper in the Second World War.

Frederick Maurice Crum ✻ (1872–1955), author of *Memoirs of a Rifleman Scout*, passed out of the Royal Military College, Sandhurst, and duly served in India and South Africa. Severely wounded during the Battle of Talana Hill, he was invalided home but subsequently returned to South Africa. A chance meeting with Robert Baden-Powell persuaded Crum to leave military service to devote time to the Boy Scout movement in Scotland. Rejoining the army when war began, service in the trenches in France once again compromised his health. In May 1916, therefore, he was sent to Acq to organise a brigade-level sniping school, which was successful enough to persuade General Skinner to move Crum and his staff to brigade headquarters in Arras to oversee the work of snipers and intelligence-gatherers. Returning to Britain in poor health, Crum was appointed 'Scouting and Sniping Expert G.H.Q. France', wrote *Scouting, Sniping and Observation* for limited circulation and, on 17 May 1917, took over co-ordination of scouting and sniper training throughout the army.

Thomas Oram Durst (1889–1964), listed as a Civil Service clerk in the 1911 census, serving as a private and then 'acting corporal' in the King's Royal Rifle Corps during the First World War. His reminiscences were used by Martin Pegler in *Sniping in the Great War*.

Fyodor Trofimovich Dyachenko ✻, born in the Ukraine in 1917, settled in the Krasnoyarsk region of Siberia when his family were deported after being denounced as 'kulaks' (suspiciously wealthy peasants). Conscripted in May 1942, Dyachenko impressed his trainers with marksmanship and, after being sent to the Leningrad Front, was given a sniper rifle. Initially facing Spanish troops sent by Franco to fight for Germany, and later Waffen-SS units, he was seriously wounded early in 1944. Said to have amassed 425 kills, principally while serving with the 187th Rifle Regiment, awarded the title Hero of the Soviet Union on 21 February 1944, Dyachenko became a political officer and served the army until retiring in 1962.

Denis Ivo Exley Edwards (1924–2008), author of *The Devil's Own Luck*, based on a diary kept after D-Day, trained as a scout-sniper with the Oxfordshire & Buckinghamshire Light Infantry. Attached to the 6th Air Landing Brigade, he was among those dropped by glider on the night of 5/6 June 1944 to capture Pegasus Bridge over the river Orne, and then took part in the thrust across Europe to Germany.

John Mark Fulcher (1913–89), one of many snipers featured in Charles Sasser and Craig Roberts's *One Shot – One Kill*, was born in Texas to German-Irish and Cherokee parents. He enlisted in the US Army in December 1942 after serving with the National Guard, and, once marksmanship training had been completed, was sent to Italy – landing at Salerno with the 36th Division in September 1943. Discharged from service in October 1944, Fulcher was the eldest of five brothers serving during the Second World War.

Harry Mitchell Furness ✻ (born in 1925), after transferring from the Home Guard in 1943 served as a sniper with the York & Lancaster Regiment, and is now credited with 117 kills. His reminiscences of service were used by Martin Pegler in *Out of Nowhere* and *Sniping in the Great War.*

John Benjamin George (1918–2009), author of *Shots Fired in Anger*, joined the US Army towards the end of 1941. A first-class shot, eventually rising to lieutenant-colonel's rank, George served as an infantry unit commander at Guadalcanal and in North Burma. His highly regarded memoir recalls experience as a rifleman in the Guadalcanal campaign.

Boris Godov, son of a priest, was born in 1920 in Ivanovo district near Moscow. Enlisting in 1935 in a 'musical regiment', Godov served in the Winter War with a Ukrainian assault brigade. Subsequently re-training as a political-education officer and first-aider, he demonstrated sufficient aptitude for shooting to be given a Tokarev rifle and sent on a sniping course. Seriously wounded in the autumn of 1941 while serving on the Ugra river, his sniping career then gave way to reconnaissance duties.

Julian Henry Francis Grenfell ✳ (1888–1915), son of William Grenfell, Baron Desborough, was commissioned into the Royal Dragoons in 1910. When the First World War began, the Dragoons were sent to northern France. On 1 January 1915, already an experienced sniper and twice mentioned in despatches, Grenfell was awarded the DSO for his gallantry. On 13 May, however, while observing enemy positions, he was struck in the head by a shell fragment and died on 26 May.

Simo Häyhä ✳ (1905–2002), a farmer's son, joined the Protective Corps in 1922 and then served the Finnish Army until 1927. When the Winter War began, Häyhä was posted to 34. Jääkärirykmentti. He is now generally credited with 505 sniper kills (some sources claim 542), but his front-line career was cut short by a Russian bullet that took away part of his jaw. Tapio Saarelainen details his remarkable story in *The White Sniper Simo Häyhä*.

Matthäus Hetzenauer ✳ (1924–2004), born in the Austrian Tyrol, was sent for training with Gebirgs-Ersatz-Bataillon 140 in September 1942, temporarily released and then recalled to service in January 1943. Graduating in July 1944 from a marksmanship course in Steiermark in Seetaler Alpen, and serving on the Carpathian Front with Gebirgs-Jäger-Regiment 144, Hetzenauer's final tally of 346 was unmatched by any other Wehrmacht sniper. Captured in May 1945, he survived captivity in the USSR to return home in 1950.

Oliver Hogue (1880–1919), 'Trooper Bluegum', was an Australian army officer who reported for the *Sydney Morning Herald* during the Gallipoli campaign. His reports were subsequently collected as *Love Letters of an Anzac* and *Trooper Bluegum at the Dardanelles*, both published in 1916. Sent to England after contracting dysentery, he died from the effects of influenza.

Albert Huxford (1897–1955), 'Gardener's Boy' according to the 1911 census, served in the Royal Berkshire Regiment during the First World War, duly qualifying as a sniper. His reminiscences were used by Martin Pegler in *Sniping in the Great War.*

Ion 'Jack' Llewellyn Idriess ∗ (1889–1979), spent his formative years in rural New South Wales. When the First World War began, Idriess joined the 5th Light Horse Regiment and was sent to Gallipoli where he acted as observer for Billy Sing. Beginning in 1927, Idriess wrote 50 books, which include *The Desert Column*, an account of his service in Gallipoli and Palestine, and a series of training manuals for the Australian Army including *Shoot to Kill* and *Lurking Death*.

Mariya Dimitrievna Kataeva (Bondarenko) ∗, born in 1925 near Moscow, was working on a collective farm when she tried to enlist as a sniper in 1942 – an effort which lasted until 1943. Graduating from the Central Women's Sniper School in March 1944, she was sent to serve with the 179th Rifle Regiment on the 1st Belorussian Front and was badly wounded several times before ending the war in Königsberg with an unverifiable tally of 28 kills (some sources claim 80) and the second and third classes of the Order of Glory.

Antonina Aleksandrovna Kotlyarova (Zakharova) gave an interview to Artem Drabkin in 2009, and also features in Andrey Ulanov's *Stalkers of the Enemy*. Serving with 47th Army after graduating from the women's sniper school, she received the Order of Glory third class, recognising ten kills, but her final score may have been considerably higher.

Nina Alekseevna Lobkovskaya ∗ was born in 1925 in the Siberian village of Fyodorovka, and volunteered for military service in April 1943. One of the first cohort of the Principal School for the Training of Female Snipers, she proved an outstanding pupil. Sent to the front with the 21st Guards Rifle Division of 3rd Shock Army, she not only ended the war with 89 kills, but was placed in command of an all-female sniper section in February 1945. Demobilised in August 1945, Nina Lobkovskaya related her sniper story in interviews given in 1997 and again in 2009.

Herbert Wesley McBride ∗ (1873–1933), author of *The Emma Gees* (1918) and *A Rifleman Went to War* (1933), enlisted in the National Guard and had attained the rank of captain by 1907. On 1 February 1915, though still a citizen of the USA, McBride was commissioned into the Canadian 43rd Infantry Regiment and attached to 21st Battalion CEF as a musketry instructor. However, his military service was characterised by drunkenness and insubordination. Private McBride, as he became after his commission had been rescinded, reached France in September 1915. After commanding machine-gunners, he demonstrated an aptitude as a sniper. Granted a temporary commission on 31 May 1916, and awarded the Military Medal in June, McBride was subsequently 'Dismissed from the Service by sentence of a General Court-Martial'. He then joined the US Army to serve as a marksmanship and sniping instructor, only to resign – once again in disgrace – in October 1918. However, as *A Rifleman Went to War* eloquently testifies, McBride was a fastidious observer and was highly regarded as a mentor by fellow sniper instructors.

Nikolay Dimitrievich Nadolko was born in the Bashkir republic in 1926, though his family were Ukrainian. An aptitude for marksmanship, learned on a collective farm, enabled him to attend a sniper school in Kazakhstan. Graduating in August 1944, Nadolko went to serve in the Ukraine with the 585th Rifle Regiment, obtaining about 15 kills before being sent, shortly before the war ended, to study in Odessa.

Aleksandra Petrovna Medvedeva (Nazarkina), listed by Alla Begunova in *Angels of Death*, served with the 508th Rifle Regiment, 174th Rifle Division. Her final tally was at least 43 kills.

Evgeniy Adrianovich Nikolaev ∗ (1920–2002), author of *Red Army Sniper* (2017) and also, with his granddaughter Dina Nikolaeva, of *Sovetskie Snayperu v Boyo 1941–1945*, a fascinating history of Soviet sniping based on newspaper articles and interviews with fellow snipers. Born in Tambov, Nikolaev worked as a theatre-set designer before being drafted in October 1940. He subsequently served with NKVD infantry units in Karelia and then at Leningrad, where he was among the first snipers to be deployed. After being seriously wounded, Nikolaev was eventually transferred to the counter-intelligence service, SMERSH; his count had reached 324 kills.

Anatoly Navara was born in 1927 near Kharkhov in eastern Ukraine. His father was a leading Communist Party official, and so Anatoly duly joined the army in 1944 to be sent for sniper training at Piryatin. Sent to serve with on the 2nd Ukranian Front in February 1944, before graduating from sniper school, Navara served for some months before returning to training, which remained incomplete when the war ended.

Klavdiya Panteleyeva (Kalugina), serving with the 344th Rifle Division of 33rd Army, registered 'at least 3' kills but is best remembered for illuminating reminiscences dictated in 2009 to Artem Drabkin and recounted in Ulanov's *Stalkers of the Enemy.*

Lyudmila Mikhaylovna Pavlichenko ✳ (1916–74), author of *Lady Death* (2018), was born Lyudmila Belova in Bila Tsverka in the Ukraine. Embarking on a five-year history course at Kiev University in 1937, Pavlichenko joined an Osoaviakhim shooting club in her teens and won many trophies. Volunteering in the summer of 1941, she was assigned to the 54th Rifle Regiment, fighting initially to defend Odessa before moving on to Sevastopol. Her first 100 kills were achieved with an Obr. 91/30 Mosin-Nagant, and though presented with a Tokarev SNT-40 at the time, she continued to use the Mosin-Nagant whenever possible. Credited with 257 kills by May 1942, Pavlichenko was then injured by a mortar-bomb fragment and withdrawn from combat to be sent with Vladimir Pchelintsev on a tour to North America and then Britain to raise awareness of the desperate struggle on the Eastern Front. When she returned to the USSR, Pavlichenko was made Hero of the Soviet Union, became a sniper instructor, and ended the war ranked as a major. She contributed a brief account of her career to a Soviet historical publication shortly before her death, and her memoirs appeared posthumously.

Vladimir Nikolaevich Pchelintsev ✳ (1919–2001) accompanied Lyudmila Pavlichenko on a tour of North America and Britain in 1942–3. He was not only among the best of the Soviet rifle shots but also an influential sniper trainer. Born in Tambov, Pchelintsev joined the army to serve on the Leningrad Front. Politically active with Komsomol, Pchelintsev was created a Hero of the Soviet Union in February 1942. His personal tally amounted to 456 – including 14 snipers – but his front-line service was terminated in favour of an educational role in 1944.

Eric Hillman Penberthy ✳ (1889–1942), author of 'British Snipers (i). An Account of the Training and Organisation of Snipers in the British Armies in France' published in September 1920 in *The English Review*, was born in New South Wales and was among those attending the Conference of Experts in Scouting, Sniping and Observation held in July 1917. Penberthy represented the British Third Army. The 1918 *Army List* records him as a lieutenant in the 'London Regiment (London Rifle Brigade)', but he subsequently transferred to the Royal Flying Corps before emigrating to South Africa.

Iosif Iosifovich Pilyushin ✳ (1903–?), subject of *Red Sniper on the Eastern Front* (2010), served the 14th and 105th Rifle Regiments of the 21st NKVD Rifle Division. Credited with about a hundred kills, though other sources give 115 or 136, Pilyushin lost an eye to a shell fragment in 1942, but re-taught himself to shoot left-handed well enough to resume his sniping career before being badly wounded and hospitalised.

Mariya Semyonovna Polivanova (1922–42), partner of Natalya Kovshova, killed on the same day, joined the army in October 1941 to serve with the Moscow Communist Rifle Division and then on the North-West Front with the 528th Rifle Regiment, 1st Shock Army. Polivanova is usually credited with only 50 kills when her total was probably higher; most sources agree on about 75. She was made Hero of the Soviet Union posthumously in 1943.

Hesketh Vernon 'Hex' Prichard * (1876–1922), son of Lieutenant Hesketh Brodrick Prichard, who had died before Hex was born. Returning to England, Hex and his mother began to write highly successful adventure fiction which allowed him to embark on a career as an adventurer-explorer. When the First World War began, Hesketh Hesketh-Prichard (as he had become known) was rejected for front-line service owing to his age, but found a niche working for the Intelligence Office. After using his telescope-sighted hunting rifle 'to help the war effort by killing some Germans', Hesketh-Prichard then attempted to impress upon senior officers that counter-sniping was practicable and find backing for his sniper-training campaign. Returning to France, he collaborated with fellow pioneer Langford Lloyd at the 'telescopic-sight school' attached to X Corps, and then went to Third Army. An excellent teacher, and a passionate advocate of the telescope sight, Hesketh-Prichard created the First Army School of Scouting, Observation and Sniping at Linghem in the summer of 1916. The school grew rapidly, employing Canadians among its staff, and attained 'provisional establishment' status in November. Innovations promoted by Hesketh-Prichard and his colleagues included use of dummy heads as sniper-locators and a way of inserting an undetectable steel-lined loophole in a sandbag wall.

Aleksandr Romanenko, born near Volgograd, won many shooting competitions in his youth – skills that earned him a place on a sniper-training course which lasted until October 1943. He was sent to the Dnepr bridgehead as part of an eight-man sniper group, was subsequently hospitalised in Kiev after being wounded, but recovered to traverse the Reich as far as Berlin when the war ended. Claiming 29 kills, though this has not been confirmed, Romanenko served the army until 1950.

Clifford Shore * (1907–56), author of *With British Snipers to the Reich* (1948), an enthusiastic marksman with no combat experience, served initially with the RAF Regiment before transferring to the army. The 1948 *Army List* notes him as captain in the Manchester Regiment (Territorial Army), but gives no other details. Shore's book is highly detailed, but often opinionated.

Roza Egorovna Shanina ✻ (1924–45), a prolific diarist, remains one of the best-known of all Russian female snipers. Her life and career are summarised in Appendix 2.

Frederick James Sleath (1889–1966), born in Borrowstounness (Bo'ness), West Lothian, graduated from Edinburgh University with a master's degree shortly before the First World War began. Commissioned in the Royal Scots, rising to the rank of lieutenant by the war's end, he then followed a career in journalism, and *Sniper Jackson*, basically a personal reminiscence in the form of a novel, was published in 1919 to be followed by five thrillers. Sleath was recorded in the 1939 Register as Journalist and Editor/Manager of the 'C. Journal'.

Bruno Sutkus ✻ (1924–2003), the subject of *Sniper Ace: From the Eastern Front to Siberia* (2009), was born in the Tannenwalde district of Ostpreussen. Transferring from the Hitlerjugend to the Sturm-Abteilung in May 1942, he moved in July 1943 to Füsilier-Ersatz-Bataillon 22 and then to the sniper school in Vilnius. Graduating at the end of December 1943, Sutkus joined Grenadier-Regiment 196 and obtained his first kill on 8 May 1944; the 50th followed on 4 September; the 100th on 16 November; and the 150th on 17 December. Reaching 200 kills on 7 January 1945, promotion to *Obergrefreiter* presaged a transfer to sniper school as an instructor. Hospitalised on 22 January after being wounded, Sutkus was captured by the Red Army in May 1945. His status as a sniper unrecognised, he survived interrogation, torture and banishment to Siberia. Bruno Sutkus re-entered Germany in 1991 and was allowed to settle there in January 1997. Remarkably, his highly detailed sniper log survives.

Richard William Tregaskis (1916–73), author of *Guadalcanal Diary*, was employed by the *Boston American Record* when the USA entered the Second World War at the end of 1941. One of only two newspapermen present on the island, his memoir, published in 1943, has been reprinted many times.

George Owen Van Orden (1906–87), son of an officer in the US Marine Corps, enlisted in the Marines in 1925 as a private and retired in 1949 ranked as brigadier-general. A well-known competition shot, winning many prizes, Van Orden was renowned for spearheading the Marine Corps Scout-Sniper School training course during the Second World War and compiling *Equipping the American Sniper* with Gunner Calvin Lloyd, while his reminiscences were also used by Martin Pegler in *Out of Nowhere*. Van Orden was also a vocal proponent of the Winchester Model 70 rifle fitted with large-magnification Unertl optical sights as an ideal sniper's weapon.

Hermann Wirnsberger achieved 64 kills while serving with 3. Gebirgs-Jager-Division from 1 September 1944 until seriously wounded. Thereafter, he trained snipers until apparently returning to combat duty shortly before the war ended in May 1945. Little else is known of his career, however.

Vasiliy Grigorevich Zaytsev * (1915–91), author of *Notes of a Russian Sniper* (2014), was born in the Ural mountains. When the Great Patriotic War began, Zaytsev, then a petty-officer clerk in the navy, stationed in Vladivostok, volunteered to serve in the front line. Transferring to the army as a senior warrant officer, and after proving his marksmanship, Zaytsev was posted to the 1047th Rifle Regiment, part of 62nd Army defending Stalingrad. Zaytsev's memoirs give a graphic impression of the war among the ruined factories and wrecked industrial installations in which attackers and defenders were often, literally, a wall's width apart. Zaytsev was particularly adept at making use of cover and became a leading advocate of changing position after each shot. While taking part in an assault on German lines in 1943, shortly after being made a Hero of the Soviet Union, he was temporarily blinded by a mortar-bomb blast but eventually returned to combat before the war ended.

Yuliya Konstantinovna Zhukova *, author of *Girl with a Sniper Rifle* (2019), was born in Uralsk in 1926. Volunteering for sniper training early in 1944, struggling successfully through eight months of training, she subsequently served on the Belorussian Front with the 611th Rifle Regiment of 88th Rifle Division. Now usually credited with at least eight kills, Yuliya Zhukova is better known for her eminently readable memoir originally published as *Devushka so Snayperskoy Vintovkoy* (2006).

RISE OF THE SNIPER

The official views of sniping, often hostile, have altered as the years have rolled past. It is probably fair to record that, the USSR/Russia excepted, the sniper has never enjoyed public esteem in a way accorded to men whose heroism was lionised.

When the First World War began, heedless of lessons learned in conflicts from the American Civil War to South Africa, the establishment view was often simply that a good target shot would make an ideal sniper (if no alternative to 'unsporting' sniping was to be found!) or that men from army marksmanship schools would simply achieve success on the front line. But this approach had its flaws, as Hesketh-Prichard explained in *Sniping in France*:

> In the [British] Army there has always been in certain quarters a prejudice against very accurate shooting, a prejudice which is quite understandable when one considers the aims and ends of musketry. While sniping is the opportunism of the rifle, musketry is its routine. It would obviously never do to diminish the depth of your beaten zone by excess of accuracy. But this war, which, whatever may be said to the contrary – and much was said to the contrary – was largely a war of specialists, changed many things, and among them the accurate shot or sniper was destined to prove his extraordinary value.

By the spring of 1915, the notion that sniping had no value was being steadily discredited. Yet, in the absence of army-wide direction (which was to happen, grudgingly, in 1917), the British achieved only individual laurels – usually gained by junior officers acting on their own initiative. The Germans still saw sniping as a 'temporary job' for qualified marksmen to whom optically sighted rifles were issued by an NCO.

The lack of an effective riposte to the German snipers was to cost the British dearly. On 5 February 1915, Lieutenant Stanley Shingleton of the Royal Field Artillery confided in his diary that he had been:

> sent in charge of a working party at night time to build dummy redoubts behind the front line, we crept along, taking advantage of the hedges etc. There were not many hours during the period between nightfall and moonrise so we had to get to work sharp for to be seen standing in the open meant instant death from a sniper's bullet. It was even dangerous to wear a luminous wrist watch for the light from the same could be seen a long way off in the pitch darkness; a few snipers' bullets whistled past us, and the R.E. officer in charge suggested that my luminous watch was the cause, whereupon I put it in my pocket!

An early and largely misplaced belief in the value of established target-shooters soon gave way to the notion that training could only be effective if snipers were drawn from those who were practised at stalking, hunting or wing-shooting.

Target-shots such as Lieutenant George Gray of the Scottish Rifles, winner of the King's Prize at Bisley in 1908 and the Caledonian Challenge Shield in 1913, often proved to be superb instructors. But Clifford Shore, writing in *With British Snipers to the Reich*, who had a poor opinion of the military value of target-shooters, observed that:

> During the training of snipers there were many instances of men being excellent target shots but failing in the role of sniper shots. When I mentioned this matter … with an officer, a very well-known Bisley shot and a sniping instructor in England and Italy during the [Second World] war, I was taken to task and my opinions rebuked in similar sentences to the following:– 'In my experience, the lad who has shot at school, or at Bisley has been streets ahead of anyone else, because he's a better shot'. Frankly, I should have thought that anyone who had sniped and worked as a sniper trainer would have found instances qualifying my contention that many range and competition shots did not make the grade … This was a point mentioned to us at the sniping school in Holland [that] some men who were very good target shots … found that shooting under service conditions was a very different proposition.
>
> Psychologically, the difference between resting on a placid firing point and shooting in war, even if one is in a comparatively safe position oneself, is the greatest possible contrast.

And Eric Penberthy recalled in *The English Review* in 1920 how:

> In the early days of the war, when reports of German 'sniping' began to be published, it was commonly considered a 'dirty' method of fighting and as not 'playing the game'.
>
> Sniping in some form or other has usually developed in every great war since the invention of firearms. Austrian sharpshooters were used against the soldiers of Frederick the Great, and were so effective that at times the ex-Kaiser's ancestor found it difficult to bring his Hussars, armed only with short sabres, out of their camps.
>
> This led to the formation in the Prussian Army of Jager battalions, recruited from hunting districts, because professional hunters had not only the necessary skill in marksmanship but also the courage, good eyesight, powers of observation, knowledge of stalking, and the use of cover which the successful sniper must possess.
>
> That is why at the beginning of the war the enemy was well equipped with expert snipers.

However, Penberthy said, the British had no one with special training until steps 'were taken in France to train and organise our snipers':

> In 1914 the circumstantial reports of the deadly work done by the Boche snipers gave the world the impression that German soldiers were better shots than we were. The German never was a better shot, or even as good a shot, as the Britisher.
>
> It was the better shooting and the superior manipulation of their rifles by the 'Old Contemptibles' that enabled the British Army to 'walk backwards' from Mons in perfect order and discipline until the enemy was effectively checked. But the shooting done by Lord French's old Army was not sniping, it was controlled fire. It was done by sections and platoons at the word of command. Targets and ranges were indicated and fire-orders given. And the result was such a torrent of well aimed and well directed bullets against masses of the enemy that captured German officers stated their belief that every British Tommy was armed with a light machine gun. But we had no snipers, men specially selected and trained to find small and indistinct targets, to shoot at them on their own individual initiative and be sure to kill at the first shot.

He then observed that 'when both sides settled down to trench warfare':

> the Boche sniper showed the value of his special training. Often behind his trench line as well as in it, and from cunningly constructed

and concealed posts, he kept a vigilant watch on our lines. He picked off sentries and observers, who carelessly or sometimes unavoidably exposed themselves.

Many officers were picked off when carrying out the dangerous but essential duty of making a daylight reconnaissance until at last the Boche sniper dominated our trenches and No Man's Land so effectively that it became urgently necessary to take steps to deal with him.

At first this form of warfare appealed most strongly to those battalions which possessed keen hunters amongst officers and men. That is why snipers were first organised most effectively in the Canadian and Highland regiments.

So the system went on in our ranks of selecting men whose pre-war occupation, or recreation, specially qualified them to be snipers. It spread from one battalion to another. If the Commanding Officer was keen on sniping, or if the sector was much bothered by enemy snipers, a dozen men might be detached for this duty. Other battalions were content with one or two.

However, despite the efforts being made, activities of German snipers were:

only slightly diminished in numbers and not suppressed. A revision of methods and further organisation became necessary, and some experienced officers were selected to study the problem and make suggestions. One of the first things noted was that a large number of men had been selected as snipers who were totally unfitted for the work. In order to carry out his duties, the sniper was allowed a great deal of freedom. He could wander about the trenches, selecting suitable posts from which to observe and shoot, and he was naturally excused a great many irksome tasks which the ordinary soldier had to do as a matter of routine.

A sniper, therefore, came to be regarded by observant but not too energetic soldiers as a man with a soft job, and there were naturally many seekers after these posts. In the trenches, the opportunities of testing a man's ability for this special work, especially in marksmanship, were very limited, and many soldiers were given the appointment who had no idea of the job.

One obvious shortcoming was the lack of appropriate training:

The first thing they had to learn was how to find their targets. It was quite a common thing in so-called 'quiet' sectors for men to be in

> the trenches for six months without seeing a Hun, and the difficulty of finding a casual head or a loop-hole can be imagined. To train these men, some keen young officer was selected in each battalion. Sometimes he was detached from his company and did nothing else, although it was a very long time before he was given the status of an officer doing special duties and allowed for on the establishment.

The key to success often lay simply in finding a 'sniping officer' who combined shooting experience with the ability to select the right kind of man and then undertake training both in the trenches and out of the line:

> As time went on it was realised that if a sniper was doing his duty properly by keeping his front under continual observation and being ready to fire on enemy snipers or observers, or anyone else who exposed himself, he must also see a host of other happenings. Although these were not directly bearing on his work as a sniper, they would, if reported to the proper quarter, provide valuable and continuous information about the enemy.
>
> Consequently, snipers were now trained to write detailed reports on what they had observed during their tour of duty. These details were recorded in a report by the Sniping Officer, who now became more generally known as the Battalion 'Intelligence' Officer.
>
> By this system of reports snipers were able to provide information about such extremely important matters as the identification of the enemy on any front, the time and method of reliefs, the presence of mining operations, and so on. Trench warfare may be said to have brought this system of collecting Intelligence into existence, and the sniper became the machinery for carrying it out.

But the situation improved rapidly once the idea of training snipers became vital. Penberthy acknowledged the forward-looking step taken when:

> In December, 1915, a school for training snipers and Intelligence officers was started near Ypres by the Second Army, with such excellent results that similar schools were soon established for all the other Armies in France.
>
> As there was no previous model on which these schools could be framed, their methods of training had to be based upon actual experience in the field and continually modified to suit new conditions. At first the shooting was considered the main thing, but later, when

> the importance of observation was realized, they became schools of 'Scouting, Observation and Sniping'.

And Frederick Crum mused in *Memoirs of a Rifleman Scout* on his last tour of duty in the Ypres Salient in February 1916:

> After six days' rest we returned to the [Lancashire Farm] trench; we found things more lively than usual, and it seemed probable that the Hun had some scheme up his sleeve. Shelling, shelling, shelling, and machine-gunning all day and night, but mostly away from our immediate sphere of usefulness.
>
> For the O.C. ['Officer Commanding'] it is watch and watch, and keep on watching and wondering what is at the back of anything unusual. For the time he is screwed up, as it were, ready to do the right thing if it may be, and able to keep cheery and confident. He must be up and about, visiting sentries and snipers and bombers and working and wiring parties, and watching through the telescope from Observation Posts, an eye on the weather-cock for gas winds.
>
> It was a great satisfaction to see the victory of our snipers. When we came away not a shot was ever fired by the Bosche sniper, and he had closed and sandbagged all his loopholes.
>
> Six weeks ago, when we took over the trenches, the Brigadier [Lord Binning] warned us that [German] sniping was bad, and so it was.

In May 1916, Frederick Crum was sent by General Skinner to create a Brigade School of Sniping at Acq. He subsequently recalled in *Memoirs of a Rifleman Scout* that the school had been a great success:

> We were full of ideas, all of us, Hicks, Harman, Sherry, Cox, Taylor and others. The scandal of hundreds of men getting bowled over, simply from want of teaching and imagination, stirred us to great efforts so that in addition to building a range in a chalk quarry close by, and greatly improving our marksmanship, we were able to give demonstrations, to troops resting out of the line.
>
> The methods of teaching are best understood from reading the small book which I wrote called *Scouts and Snipers in Trench Warfare.*

In a subsequent, somewhat critical, 'Report on Scouts and Snipers', Frederick Crum related how the situation on the Western Front remained problematical even as late as May 1917:

> We completed another … course and I had sent in the following Report, which led to my being appointed as Scouting and Sniping Expert at G.H.Q. in France.
>
> PRESENT POSITION. While we constantly miss chances of inflicting losses in men and morale on the enemy, we ourselves constantly incur avoidable losses. We have no recognised system, or organisation, or training, and no one at the head to secure co-operation. The Schools at Home and in France are not in touch and vary. Men trained as Scouts are often not used as Scouts, chiefly because many Officers are ignorant of their uses and value, and may even be prejudiced against Scouts.
>
> CO-OPERATION. Thus there is no co-operation between Scouts of neighbouring Battalions, or of Battalions relieving each other in the Line; nor do Scouts work as they should in touch with sentries, or with Trench-mortar, Artillery, and Intelligence Officers …
>
> WASTE. There is considerable waste of valuable material, such as telescopic-rifles, periscopes, sniperscopes, dark-glasses, etc., issued often to men untrained in their uses; at the same time it is often difficult to obtain these articles, both for training and for use at the front.
>
> LITERATURE. No official guide is yet available. Divisional Generals, Brigadiers, and keen young Officers are constantly enquiring for guidance. I forward herewith a printed booklet which has been commended by G.H.Q. France, and I could be glad if anything helpful in it could be made available. Some such system and training with model German and British trenches; with teaching by acting, training in the use of telescopic rifles, dark-glasses, jiu-jitsu, camouflage, etc., would give valuable results.
>
> SUGGESTIONS. At present we are like some valuable war-ship without chart, course, compass, rudder or Captain. I suggest that some suitable Officer should be appointed, with Staff and authority, to deal with the matter. That he should hold a Conference of Officers in charge of Schools, and that they should submit some sound scheme an Official handbook to follow.

Crum also stated that, from May to November 1917, the work with which he had been entrusted in France:

> was to carry out the 'co-ordination' for which I had always been agitating. There were (at the time) five schools of Sniping, one in each

> of the five armies; each of these Schools was working on its own lines, some devoting special attention to the scientific side of shooting and telescopic rifles, some making a speciality of observation, and some of camouflage. The wish of the Training Authorities was to introduce more night patrol work and open warfare scouting and sniping.
>
> The days of trench warfare seemed numbered and it was not considered advisable to keep a special staff merely for training in scientific marksmanship. The tendency at that time was to do away with the specialist and make men all-round handymen. How far it was possible to turn out men efficient both in sniping and scouting, were questions on which experts did not agree, any more than they did when it was laid down that an Infantry soldier must be efficient in bombing and Lewis Gun as well as with bayonet, rifle, and the rifle grenade!

However, there were still many problems to be overcome:

> The 'powers that be' at G.H.Q. at the time were not in favour of sniping at all as a separate art; they were set on further training in Scouting, in which we had been found sadly deficient in those glimpses of open warfare afforded on the Somme, at Wytschaete Hill and further North. Thus at the outset one had to find out what was possible, and then how best to secure the soundest compromise in training.
>
> One difficulty was that the authority of the Training Staff at G.H.Q., being newly established, old established Army Schools and Corps and other schools of all kinds often thought they knew better than G.H.Q. Possibly they did, sometimes, anyway. The process of co-ordination of all Education and Training was then at an early stage.

Crum had spent two months visiting army schools, corps and divisional headquarters, with occasional visits to brigades and battalions along the front:

> In this way one came into touch with the views of all kinds and conditions of men. The next step was to hold a Conference of Experts in Scouting, Sniping and Observation at Boulogne, where the various views were discussed.
>
> Thirdly, at Bouchon, after training a dozen young soldiers with the help of expert officers and N.C.O.'s, I was able to work out and demonstrate to representatives of each Army H.Q. and Sniping School a suitable system of intensive training in Scouting.

> Lastly with the help of other experts and the experience gained, my time was taken up with the completion of an official book on 'Scouting, Sniping and Observation,' which became known as 'SS 195'. This was the first official book ever produced by the Army on the subject and is therefore of interest.

However, not everyone agreed with Crum's often self-promotional views. For example, Herbert McBride observed in *A Rifleman Went to War*:

> Impossible though it may seem, it is possible for a cool, resourceful man to circumvent and actually destroy a large number of the enemy. But, leaving aside these brilliant exploits, the rifleman in battle has his own work cut out for him. It is not easy; not by a long shot. Nothing is easy when a man is stumbling and crawling and creeping through an inferno of flying, howling and shrieking missiles: through dust or mud, straining every muscle to move at all – probably wearing a gas mask which will fog up and sweat, in spite of all the dope that can be put on it. No, no, it is not easy.

Snipers and Gallipoli

The attack on the Dardanelles was an unmitigated disaster. The Allied commanders greatly underestimated the abilities of the Turkish Army, honed by the bloody Balkan wars and bolstered by German advisers. By nightfall on 25 April 1915, invasion day, 2,000 of the 16,000 men of the 1st Australian Brigade landed at Gallipoli had become casualties.

Making good use of cover, firing from undetected pits close enough to the Allied lines for even parapet loopholes to be hit far too easily, many of the Turks were local men: experienced hunters, at home in the terrain, and well-practised shots. Usually equipped with an open-sighted 7.65mm M1893 Mauser, they customarily had the advantage of high ground. Colonel Talbot Hobbs, commanding the Australian artillerymen, remarked that the snipers:

> well supplied with ammunition and food, and frequently camouflaged with Hessian painting to assimilate with the scrub, watched their opportunity to pick off our men, and it was some considerable time before reports ceased to be sent into various headquarters that snipers were still inside our lines.

Casualties from sniping escalated alarmingly. Lieutenant Colonel Lancelot Clarke, commanding 12th Battalion Australian Imperial Force (AIF), had been killed on the beach on 25 April; Colonel Henry MacLaurin, commander of 1st Brigade, was struck down two days later; and Major General Sir William Throsby Bridges, commanding 1st Division, was shot while inspecting a defensive position on 15 May 1915 – a wound which proved fatal three days later.

Charles Bean recorded in his *Official History of Australia in the War of 1914–1918* that, in the first few weeks at Anzac Cove, Turkish riflemen managed to penetrate the Australian and New Zealand positions after every attack. During a strong enemy assault on 1 May, an enemy party had:

> crept along the side of Russell's Top into the large bare watercourse … in the heart of the Australian position. Individual snipers probably stole even further, and men on the slopes in the rear of the line in Monash Valley were still shot dead from behind. The snipers were sometimes found and killed. As late as May 1st parties were sent to clear them out with the bayonet. But even with Russell's Top free of them, the Turks on The Nek and the Chessboard still looked into the back of the line from Quinn's to MacLaurin's Hill, while those upon Dead Man's Ridge stared at point-blank range down the valley.

By late-summer 1915, sniping and counter-sniping had become of such vital importance at Gallipoli to be incorporated in tactical plans and were often co-ordinated with flank attacks. Snipers acted in pairs, relieved of other duties to allow them to concentrate on their task. Anzac riflemen were often excellent shots – many had been professional kangaroo hunters – but, at least at the outset, would simply blaze away at the Turks from their strongpoints.

Eventually, a 'blind eye' turned by commanders to counter-sniping allowed anyone to take on the Turkish marksmen: cooks, clerks, even chaplain Andrew Gillison. Sometimes these initiatives were successful, but no real progress was made until Major William Malone, commanding the Wellington Infantry Battalion, decided to take action.

Malone's 'picked men' included Jesse Wallingford – born in Woolwich, Champion of Europe in 1899 and part of the British team at the 1908 Olympic Games. Wallingford had been serving as an instructor at the Hythe School of Musketry when he accepted a commission in the New Zealand Army in 1911. Lieutenant Nathaniel Swinard, born in Otago in 1889, had been Queen's Rifle Champion in 1910. And Waikato-born Second Lieutenant Thomas Grace was known by some as the 'Sniping King'.

Wallingford was duly charged with teaching men how to use the newly introduced periscope rifles, while Swinard and Grace reported to Colonel Harry Chauvel of the 1st Australian Light Horse Brigade. Chauvel then instructed Grace to raise a sniping section from the Australians.

A well-respected marksman, Grace was ideally suited to his new role and there was no shortage of recruits. Farmers and hunters steeped in the wild country that comprised much of Australia and New Zealand were often expert at tracking and snap-shooting at fleeting targets. They began to take such a toll of Turks that the Australian commander ordered his men to follow the example of the New Zealanders, and, eventually, a hundred Anzac snipers were transferred to the British lines to ply their trade there. Grace himself was mentioned in despatches in November 1915, but by then was dead: killed, probably by a Turkish sniper, on 8 August.

Charles Bean described how, moving into hides well before daybreak, watching and waiting for a suitable target to appear, men formed close bonds:

> Each observer lay all day with his telescope, and beside him the sniper with his rifle, watching some small area of the enemy's trench or territory so closely that they could eventually mark the least alteration in it ... Within a few weeks the enemy's sniping had been so completely suppressed that traffic in the valley went uninterrupted throughout the day ... Grace's snipers, posted throughout the valley, placed a barrier as impenetrable as any earthwork between the traffic in Monash Valley and the Turks whose trenches overlooked it. Thenceforward, provided the snipers were first warned, even a convoy of mules could go to the supply depot near the head of the gully at midday without a shot being fired at it. The conditions were so reversed that, whereas the track had previously been dangerous only by day and safe by night, it now became safe by day and dangerous only after dark; for the enemy's snipers, silenced during daylight, now laid their rifles on to certain points in the hope of catching wayfarers at night.

Among the Anzac marksmen were New Zealander Richard Travis, who had enlisted in September 1914 in the 7th (Southland) Mounted Rifles. After proving his skills at Gallipoli, Travis reached the Western Front in April 1916 where his scouting, sniping and counter-sniping skills were put to good use once again.

Australians Percy Thurlow and Reginald Ilett of the AIF's 22nd Infantry Battalion, respectively a blacksmith and a dairyman in peacetime, had both volunteered in the spring of 1915 and were to be awarded the Military Medal for their services. Thurlow, who had won the King's Medal in 1914, was a long-time member of the Railway Rifle Club. His Lee–Enfield was fitted with a privately purchased Aldis telescope sight. However, the most famous marksman was William 'Billy' Sing, born in 1886 to a Chinese father and English mother, who became known as 'The Assassin' or 'The Murderer'.

A successful kangaroo hunter, Sing had enlisted on 24 October 1914 in the 5th Light Horse Regiment and, after basic training in Egypt, was sent to Gallipoli. When experienced marksmen were sought for counter-sniping duties, Sing – his abilities well known – began his sniping career from Chatham's Post, firing a standard SMLE (the short version of the Lee–Enfield magazine rifle) which may have been fitted at one time with a micro-adjustable competition back sight.

Taking cold-blooded revenge on the Turks, helped by his spotters Tom Sheehan and Ion 'Jack' Idriess, Billy Sing was being credited with 119 kills as early as September 1915 by Brigadier Granville Ryrie, commander of the 2nd Australian Light Horse Brigade. On 23 October, General Sir William Birdwood, commander-in-chief of the AIF, noted 201 kills 'including unconfirmed', and Major Stephen Midgeley, Sing's controller, put the total as 'about 300' when Allied forces withdrew from Gallipoli in January 1916.

Uneasy Peace, Certain War

One of the strangest features of the post-Armistice era was the regularity with which sniping attracted disapproval, most armies clearly believing that problems would be overcome when needed simply by selecting the best shots from the rank-and-file.

The British government, under constant pressure to scale-down military expenditure, ceased to regard sniping as essential to the establishment. Scouts were retained, but no longer given specialist training. In addition, pre-1918 snipers were generally reluctant to acknowledge what they had done and simply walked away in the hope – often dashed – of resuming normal lives.

The British withdrew their sniper rifles, telescope sights and associated equipment in the early 1920s, and the Germans followed the lead in the 1930s. Nor was enthusiasm perpetuated in France or in the USA. Consequently, when they first became embroiled in hostilities in 1939 and 1941 respectively, the British and American forces paid only cursory attention to sniping until fortunes of war dictated otherwise. For the British, the period that followed the Dunkirk evacuation was the nadir. Clifford Shore wrote in *With British Snipers to the Reich* that:

> After Dunkirk, sniping was thought to be completely out and the sole cry was for more dive bombers, tanks, anti-tank guns, artillery and fire-power generally. And so, although snipers equipped with the P 14 rifle and the 1918 pattern 'scope continued to exist on paper, there was no sniper training carried out at all. After Tunisia and Sicily, however, the position was totally changed, and one of the dominant cries was 'Sniping'.

The Russian Story

Only in the USSR was large-scale marksmanship training encouraged in the period between the wars, a view which was to be vindicated by the outstanding success of Russian snipers, men and women alike, on the Eastern Front.

The Winter War with Finland was the prelude to the Great Patriotic War. When the Finns refused to cede territory, the Russian faked an artillery bombardment of their own territory, then invaded on 30 November 1939.

By the time peace was agreed on 12 March 1940, the Russians had learned some hard lessons. A reassessment made in the 1990s by a team led by Colonel-General G.F. Krivosheev acknowledged no fewer than 126,875 Russian deaths (including 36,369 men lost without trace).

One of the enduring stories of the Winter War is that of sniper Simo Häyhä, who had been credited with at least 505 kills – some sources claim 542 – in a front-line career of less than a hundred days. There is no doubt that the Russian soldiers were badly led, poorly trained and lacking clothing and equipment suitable for use in arctic conditions, often making themselves comparatively easy targets, but Häyhä's success was unparalleled. His skills, which were to have a galvanising effect on the Red Army, were soon apparent.

Almost as soon as he had reached the front line, Häyhä had been allocated a specific task. He subsequently related how his commanding officer:

> Lt Juutilainen, 'the Horror of Morocco' as he was known from his previous service in the Foreign Legion, tried to kill an enemy sniper with a scoped rifle. This Russian had taken up position about 400 meters from us and was constantly shooting towards our lines. After a while, the lieutenant sent for me and showed me approximately where he thought the enemy sniper's position to be. One of our second lieutenants was with us, acting as a spotter, when our duel begun. At first, I did not see a trace of him, just a small rock where he was supposed to be. After careful investigation, we spotted him behind a

little hump of snow near that rock. I took a careful aim with my trusty M/28–30 and the very first shot hit the intended target.

When the Germans invaded the USSR on 22 June 1941, they made rapid progress against the badly led, badly trained and badly equipped Russians. Soviet commanders, initially reluctantly, then backed anything from partisan warfare to in-service sniper training.

In 1931, the Communist Party hierarchy had ordered that basic military training, including marksmanship, should be mandatory for all boys and girls who entered elementary school, and efforts were made to recruit men and women who had trained with the Soviet youth league (Komsomol).

Credit for giving impetus to sniping in the Red Army has been given to Feodosiy Smolyachkov, who began a people's movement in Leningrad in the autumn of 1941. Credited with 125 kills from 126 'combat shots', to universal disbelief elsewhere, Smolyachkov did not live to see the effects his inspiration had on the course of the war. And, unfortunately, the protracted Leningrad campaign was to be airbrushed from history by Stalin, who wished to promote Stalingrad as saviour of the Motherland.

In the beginning, the Russians trained men for a week, then sent them to the front in the charge of an experienced sniper. However, though men graduated from these short-term courses in large numbers, life expectancy could be short.

Vasiliy Zaytsev claimed that 80 per cent of trainees, *Malenkiy zaychik* ('little hares'), lasted less than two weeks, but mistakenly relying on survival of the fittest to improve results soon gave way to something better and the first attempts to train snipers in large numbers were made.

There was no shortage of recruits as Osoaviakhim (the Soviet volunteer defence organisation) and Vsevobuch (the Universal Military Training programme) had prepared millions for the role. Early in 1942, Stalin ordered marksmanship to be put on a proper footing. Consequently, the Principal School for the Training of Sniper Instructors was created on 20 March. Opened officially on 3 May, and renamed the Central Sniper Instructor School on 27 November, it had actually accepted its first draft on 18 April 1942.

Attendees, volunteers and conscripts alike, had to be at least 20 years old, physically fit, and to have completed not only seven years of schooling but also a minimum of 110 hours on Vsevobuch or comparable marksmanship courses. Some came from the ranks of the Voroshilov Marksmen, but there were many from remote districts where survival depended on hunting. One such hunter was Maksim Passar (1922–43, 236 or 237 kills), an ethic Nanai from Nizhny Katar in the Khabarovsk area of Siberia, who became a largely unsung hero of the battle for Stalingrad.

The tallies of the most successful snipers grew to staggering numbers: Mikhail Surkov of the 4th Rifle Division, for example, has been credited with 702 kills, while Vladimir Salbiev of the 71st and 95th Guards Rifle Divisions claimed 601 to 31 December 1944. There is no good reason to suspect the validity of such high totals – Simo Häyhä had allegedly achieved 505 in less than a hundred days during the Winter War of 1939–40 – but there is very little concrete evidence to support some of them, and the Soviet propaganda machine played a part in the numbers game.

Ivan Sidorenko, with 500 kills, is still sometimes considered to be the true 'top scorer': the implication being that those above him had their kills inflated either by involvement in submachine-gun assaults or by political masters intent on eclipsing the detested Finn, Häyhä, who had taken such a toll of Soviet troops during the Winter War. Unusually among the Red Army snipers, who were often drawn from the ranks, Sidorenko already ranked as a junior lieutenant when the Germans invaded the USSR. He proved to be not only an excellent shot but also a first-class teacher and was charged with organising sniping courses at a time when most of the high-scoring snipers had hardly begun their service careers.

Eventually, after being wounded for the fourth time, Sidorenko was ordered from the front line to concentrate on training and is credited with the development of techniques which are still used by the Russian Army.

Vasiliy Zaytsev's memoirs give a graphic impression of the horror of war among the devastated landscape of ruined factories and wrecked industrial installations that characterised Stalingrad, a battle in which the German attackers and the Soviet defenders were all too often within range of a thrown grenade, or literally on the other side of the same wall. Zaytsev soon proved his value not merely as a sniper but as a teacher. He was particularly adept at making use of cover and changing position after each shot. Among his innovations was the use of 'sixes' – three two-man teams, one central and two flanking – to ensure that each target area was always covered from different angles.

It is regrettable that so few memoirs of the Russian snipers survive, as Zaytsev's active career, exceptionally successful but comparatively short and centred largely on Stalingrad, is atypical. The memoirs of Evgeniy Nikolaev of the 14th NKVD Rifle Regiment (324 kills, 'including a general'), published in English in 2017 as *Red Army Sniper*, present a much broader view. Nikolaev and his granddaughter Dina also compiled a multi-faceted history of sniping, *Sovetskie Snayperu v Boyo 1941–1945* (2009).

Sniping and Women

The employment of women as snipers has been subjected to great scrutiny. Though the Russians made extensive use of them during the Second World War, claims have been made of earlier use too – for example, that Turkish women faced Anzac forces at Gallipoli.

War artist Norman Wilkinson, author of *The Dardanelles: Colour Sketches from Gallipoli*, made an uncorroborated claim that Turkish snipers included 'men and even women behind our firing line'. Diaries kept by civilian contractor William Small and Private Errol Devlin of 18th Battalion AIF are among those that mention female snipers, and a letter written in June 1915 by New Zealander Reginald Stevens, of the 2nd Canterbury Brigade, states that:

> We got a female sniper, but she was shot before we knew she was a woman. There have been a lot of female snipers about. They are good shots.

No evidence has ever been produced in support of these stories, though wives, sisters and daughters of local men killed while serving the Turkish Army may very occasionally have taken up arms in search of revenge. Similarly fanciful claims were to appear after 1939, as Clifford Shore noted in *With British Snipers to the Reich*:

> During the early days in Normandy there was a terrific amount of nonsense printed in the newspapers about sniping. A frequent comment was that there had been a great number of snipers left behind on the initial German withdrawal from the coast and that these were causing much trouble behind the British front; again, quite a number of women, probably the French wives of the German occupational troops, possessed rifles and were sniping.

As the Second World War progressed, in ever-increasing desperation, the Soviets sent women to the front lines in large numbers. Most of them served as nurses, telegraphers or typists, but some served as aircrew, including pilots, or with the infantry as machine-gunners.

The Red Army's women snipers were greatly feared by the Germans, gaining the nickname *Flintenweibe* ('shotgun women'). Considered as a group, they were extremely efficient – partly due to training, for which requirements included the mandatory completion of several grades of education.

Lyudmila Pavlichenko, ultimately credited with 309 kills and the author of *Lady Death*, is the best known of them. Born in the Ukraine in 1916, she joined an Osoaviakhim shooting club in her teens and began studying history at Kiev University in 1937. The Germans had invaded the USSR before she could complete her course, but Pavlichenko soon volunteered for combat – only to encounter a sceptical recruiting officer. She recollected:

> Those born between 1905 and 1918 and liable for military service were subject to conscription in the army. Being born in 1916, I fell into this category. Having no doubt that they would take me on the spot and with pleasure, I set off for the military commissariat of the Odessa water-transport district. My meeting with the enlistment office seemed to me to be a very formal one, and I put on my best crêpe-de-chine dress and some beautiful white high-heeled sandals. In my handbag I had my passport, student card and graduation certificate from Kiev's Osoaviakhim sniper school.

But her preparations proved to be in vain:

> The room was stuffy and filled with smoke. There were doors slamming every minute. A hoarse military registrar with a face of red and blue blotches was trying to explain something to two lads of rustic appearance who had approached him. He looked at me with a harassed expression and said: 'Medical staff will be enlisted from tomorrow'.
>
> 'I'm not a medic', I replied, but he immediately turned away, indicating that our conversation was over. However, I was not prepared to accept that and placed my sniper's certificate on the desk in front of him. The irritated registrar said that there was no mention in his list of a 'sniper' category. Then he added a sarcastic comment about … women who wanted to be soldiers but had no idea how difficult it was.

But Pavlichenko was not going to give up easily, concluding on reflection that, owing to her prior shooting experience:

> My name would be entered in the list of snipers in the enlistment office in the Pechorsk region of Kiev. The previous year I had successfully competed in the city shooting competitions and passed a re-training course. Perhaps they were looking for me in the Ukrainian capital, but here I was on the shores of the Black Sea. I needed to ask the local enlistment office to make a call to Kiev.

> The following day I set off again for the military commissariat of the water-transport district. The registrar greeted me much more amiably. Judging by his reaction, he already knew what a sniper was. He leafed through my passport … and asked if my husband had any objection to my volunteering for the Red Army. I had not seen Alexei Pavlichenko for three years and replied that he had no objections.

Even after acceptance as a sniper, she encountered prejudice. In *Lady Death*, Pavlichenko explains how:

> Our regimental commander … requested that I quickly implement an accelerated programme to train a group of riflemen capable of hitting the enemy without fail. In response to my claim that it was impossible to do this in three or four days, Major Matusyevich generously gave me a week …
>
> However, it would not be fair to omit one particular feature of my experience of being in charge. The future riflemen (from other companies) did not at first know that Sergeant Pavlichenko was a woman, and they reacted to my first appearance in front of the squad in a way that was, to be blunt, unorthodox. Quite sharply, but without resorting to unprintable abuse, I took them in hand and subsequently employed very stern methods with the lazy, the slovenly and the not so bright. I had army discipline and order on my side. I soon convinced my subordinates that folk proverbs such as 'A chicken's not a game bird and a woman's not a man', 'The woman's path – from the door to the hearth', 'Long of hair, but short of brain', and other aspersions would have no currency here. I shot better than all of them, I knew a lot about the war, and they had to submit to me unquestioningly.

Private Pavlichenko was assigned to the 25th Rifle Division, defending Odessa, before moving to Sevastopol. By May 1942, Lieutenant Pavlichenko was being credited with 257 kills; several weeks later, however, a mortar-bomb fragment injured her leg so badly that she was withdrawn from combat.

Later in 1942, with Nikolay Krasavchenko of Komsomol and fellow sniper Vladimir Pchelintsev, Lyudmila Pavlichenko toured North America and Britain in a bid to raise awareness of the Eastern Front. When she returned to the USSR, Pavlichenko was made a Hero of the Soviet Union. Deemed to be too valuable to risk in combat, she was forced to become a sniper instructor. Ending the war ranked as major, Pavlichenko became a research assistant to the supreme headquarters of the navy, and died in 1974.

More than 50 Soviet women have been credited with at least 50 kills, and some high-scorers had served – and sometimes died – before their roles were formalised. Elizaveta Mironova, for example, serving with the 255th Marine Brigade, died of wounds on 29 September with a tally reckoned to be about 100 kills. Natalya Kovshova and Mariya Polivanova, a highly successful sniper team credited with up to 167 and 75 kills respectively, also died before the sniper school had been created: blowing themselves up on 14 August 1942 to avoid capture, and affording propagandists the opportunity to highlight the heroic performance of 'mere girls'.

No distinction was drawn between the sexes until a three-month training course was organised specifically for the women on 7 December 1942. There were many doubts that this was worthwhile, Oleg Savin going so far as to suggest in the 'Girls for the Front' chapter contributed to the Nikolaevs' book on sniping in the USSR in 1941–5, that:

> Fighting is not a pursuit for women. And, no doubt about it, it was not easy for the girls, who had exchanged cotton dresses for green tunics, to adjust to strict military regulations.

However, Savin continued:

> They resolutely mastered the parade-ground and bayonet drill, completed forced marches of 30–40 kilometres with full kit (rolled greatcoat, cumbersome artificial leather *kirza* [textile-based artificial leather] boots, hand-grenades, sapper's spade, rifle, helmet, gas-mask and cartridges), and learned how to crawl on their elbows, close to the ground, and if you made a mistake, you had to go back and start again. Their commanders, the majority of whom were officers who had been wounded at the front and left hospital early, taught the girls how to entrench themselves; in two hours a pair of snipers in a field or at the edge of a forest had to dig a battle trench and thoroughly camouflage it with turf, so that the enemy's entire defence system, all his trenches and all his dug-in firing points were visible.
>
> The girls … steadfastly studied the art of sharp-shooting, the intricacies of ballistics, sniper tactics, the skills of finding one's bearings in any situation and swiftly and inconspicuously switching one's targets and positions. They were taught to shoot from a sound, from the sight of smoke, to hit several targets in a matter of seconds. They were given six cartridges and, in an instant the future snipers would seek out and destroy a machine gun camouflaged in greenery

> 800 metres away, or the glass of a periscope or a target in a gun-port which would appear for only a few seconds. They learned how to throw a hand-grenade at a target and to master not just rifles, but anti-tank guns, and both light and medium machine guns.

The results were so encouraging that the Principal School for the Training of Female Snipers opened in Vechinyaki on 21 May 1943. The upper age limit was raised from 20 to 25, educational requirements were refined, and the courses lasted up to seven months.

Moving on 5 June 1943 to Amerovo in the Chelkovskaya district and then to Silikatnaya near Podolsk, what had become the Central Women's Sniping School trained three drafts for service – sending 104 women to the fronts on 22 June 1943, 585 on 24 January 1944, and another 559 in November 1944. Many of the hundreds of trainees who had been retained after graduation to serve as instructors also saw combat. But the fourth draft was aborted when it was clear that the war was coming to an end, and the school itself closed on 10 May 1945.

Graduates of the sniper school went on to achieve 11,280 kills, according to Vera Murmantseva in *Ratni i trudovi podvig sovetskikh zhenshchin*. Nina Lobkovskaya had even risen to command a sniper company serving 3rd Shock Army, about 50 strong, which alone has been credited with 3,112 kills. Losses had also been high, and it has even been suggested, albeit on doubtful authority, that three-quarters of the sniper-school graduates had been killed or incapacitated by the war's end.

Evidence is lacking, although of six women shown in one well-known photograph of Roza Shanina and colleagues in Lithuanian costume, taken on 11 September 1944, only two survived. Roza also remarked in her diary, without distinguishing between dead and wounded, that 'in our second battalion there are only 6 [snipers] left out of 78', and Yuliya Zhukova noted, possibly referring only to her cohort, that 'more than 250' of her sniper-school graduates had been killed and it is very likely that many more had been injured.

Showdown in North Africa and the Mediterranean

Australian, British, German and possibly even a few Italian snipers engaged in battles fought across North Africa, until Axis forces withdrew after the success of the British Eighth Army at El-Alamein and the recapture of Tobruk. The desert was far from ideal: cover was minimal, while the sun's glare and heat haze often distorted the visibility sufficiently to inhibit marksmanship.

Yet the third volume of the *Official History of Australia in the War*, 'Tobruk and El Alamein', contains a testimonial made on 7 June 1941 by Major Ballerstedt, commanding a battalion of German mechanised infantry which had held the line opposite the 2/15th Battalion at Tobruk:

> The Australian is unquestionably superior to the German soldier in the use of individual weapons, especially as snipers; in the use of ground camouflage; in his gift of observation; in every means of taking us by surprise …
>
> Enemy snipers have astounding results. They shoot at anything they recognise. Several N.C.O.s of the battalion have been shot through the head with the first shot while making observation in the front line. Protruding sights in gun directors have been shot off, observation slits and loopholes have been fired on, and hit, as soon as they were seen to be in use.

It has been claimed that the Australian marksmen used No. 3 Mk I (T) rifles, though evidence remains circumstantial – and one report reveals that Warrant Officer Mitchell and Private Mulgrave of 2/11th Battalion not only captured a Mauser rifle fitted with an optical sight but also formed a sniping party and were able to obtain more than 20 hits at ranges up to 900 yards.

However, in the aftermath of the Italian and then German invasion of Greece, the assault by German airborne forces on Crete in May 1941, *Unternehmen Merkur*, soon led to the hasty evacuation of British, Australian, New Zealander and Greek forces – a battle which gained Alfred Hulme of the 23rd New Zealand Battalion one of only two Victoria Crosses ever to be awarded directly to snipers.

Operating largely on his own, Hulme successfully stalked and eliminated several German snipers, acquiring not only a *Tarnjacke* (camouflage smock) but also a Zf.-Kar. 98k which he promptly turned against its former owners. When the 23rd Battalion began to retreat, Hulme stayed behind to snipe at advancing patrols. Eventually regaining his lines, Alfred Hulme was evacuated to Sphakia and then to Egypt, where his company commander and Brigadier James Hargest, commanding the 5th Infantry Brigade, duly recommended Hulme for Britain's highest valour award.

Published on 14 October 1941 in the *Supplement to the London Gazette*, no. 35306, the citation read:

> Serjeant Hulme exhibited most outstanding and inspiring qualities of leadership, initiative, skill, endurance, and most conspicuous

gallantry and devotion to duty from the commencement of the heavy fighting in Crete, on 20th May, 1941, until he was wounded in action 28th May, 1941. On ground overlooking Maleme Aerodrome on 20th and 21st May he personally led parties of his men from the area held by the forward position and destroyed enemy organised parties who had established themselves out in front of our position, from which they brought heavy rifle, machine-gun and mortar fire to bear on our defensive posts.

Numerous snipers in the area were dealt with by Serjeant Hulme personally; 130 dead were counted here. On 22nd, 23rd and 24th May, Serjeant Hulme was continually going out alone or with one or two men and destroying enemy snipers. On 25th May, when Serjeant Hulme had rejoined his battalion, this unit counter-attacked Galatos Village.

The attack was partially held up by a large party of the enemy holding the school, from which they were inflicting heavy casualties on our troops. Serjeant Hulme went forward alone, threw grenades into the school and so disorganised the defence, that the counter-attack was able to proceed successfully.

On Tuesday, 27th May, when our troops were holding a defensive line at Suda Bay during the final retirement, five enemy snipers had worked into position on the hillside overlooking the flank of the Battalion line. Serjeant Hulme volunteered to deal with the situation, and stalked and killed the snipers in turn. He continued similar work successfully through the day.

On 28th May at Stylos, when an enemy heavy mortar was severely bombing a very important ridge held by the Battalion rearguard troops, inflicting severe casualties, Serjeant Hulme, on his own initiative, penetrated the enemy lines, killed the mortar crew of four, put the mortar out of action, and thus very materially assisted the withdrawal of the main body through Stylos. From the enemy mortar position he then worked to the left flank and killed three snipers who were causing concern to the rearguard. This made his score of enemy snipers 33 stalked and shot.

Shortly afterwards Serjeant Hulme was severely wounded in the shoulder while stalking another sniper. When ordered to the rear, in spite of his wound, he directed traffic under fire and organised stragglers of various units.

Up the Beach and into the Jungle

The Japanese attack on Pearl Harbor on 7 December 1941 and the immediate declaration of war put the US Army in the embarrassing position of facing a major conflict with no suitable rifle for its marksmen. In his report to the Director, Plans and Policies, United States Marine Corps, dated 8 April 1941, the President of the Equipment Board wrote that:

> No special training program for snipers is contemplated by the army and no steps are being taken to procure special equipment. Special equipment is costly and requires special personnel to maintain and use it. Procurement problems at this time and supply problems in case of active operations would be difficult.

In fact, even when the US Army took part in the invasion of Europe – first at Salerno in 1943 and then in Normandy on D-Day – no proper sniper-training schedule had ever been drawn up. Charles Sasser and Craig Roberts noted in *One Shot – One Kill* that:

> Out of necessity, Americans initiated their own sniper programs to counter Japanese marksmen in the Pacific and Germans in Europe. Little uniformity existed in these programs in how snipers were selected and trained, even from battalion to battalion. While a number of sniper schools were set up, both stateside and in occupied territory, some gave five weeks of intensive training and others offered a day's familiarization before shipping the 'snipers' to the front.
>
> Many units made no pretense of sniper training; they merely selected those who shot the highest scores on the rifle range and dubbed them with the title.

The US Marine Corps (USMC) were only marginally better equipped than their Army counterparts, and they were woefully short of the skills needed to fight a long sniping war. Instructional courses undertaken between the wars concentrated on marksmanship at the expense of field training, but the Corps had soon been forced to realise that accurate shooting with a rifle was not enough in itself to prepare a sniper for war.

Fortunately, the Corps could call on men such as George Van Orden, Claude Harris, Calvin Lloyd and John Smith. Like Frederick Crum, Hesketh-Prichard and Herbert McBride, these men were sniping enthusiasts with enough influence to create training programmes. Writing shortly before the

Japanese attack on Pearl Harbor occurred, Van Orden had observed that the value of the battlefield sniper was:

> not to send a hail of rapid fire into a squad or company, it is to pick off with one well-directed, rapidly delivered shot, a single enemy. He must harass the foe, hammer relentlessly upon the nerves of the rank and file of the opposing forces, until his rifle 'crack' becomes a menace feared more than the shrieking shells or explosive hail from mortars. His bullets must come from nowhere.

Australian, British, Indian and New Zealander units all took part in the south-east Asian campaigns, notably in Burma, while assisting the US-led drive across the Pacific.

For many Allied snipers, the learning curve proved to be steep. Australian Sergeant Trevor King, in Glenn Wahlert and Russell Linwood's *One Shot Kills*, warned of the dangers of Japanese snipers:

> If you had an idea that there was a sniper high up in a tree, say to your left, and you're looking up for him, that's the worst thing you could do, because as you raise your white face it shows and it's picked out in a flash and fired on. So you made sure that you looked at him through the corner of your eye or underneath your hat, or steel helmet.

In addition, absence of specialist equipment, at the outset at least, was often a handicap. Though some of the Australians had the heavy-barrel version of the SMLE, many relied on No. 3 Mk I (T) rifles which had been brought out of store for refurbishment pending the availability of the No. 4 Mk I (T). Yet the old rifles often gave good results. Private Donald Latimer of 2/7th Independent Company had been given a No. 3 Mk I (T) with a P/18 telescope sight, which he liked very much as:

> It was a good, accurate rifle with a Mauser-type [action] … and a smooth two-stage trigger, but it seemed long and heavy after the SMLE. It was a bloody nuisance in the jungle … Once we were in action I usually left the rifle with the 'Q' at platoon base camp as in our patrol actions we needed Owens [submachine guns] more for fire power in our contacts …
>
> Sometimes the patrol leader would tell me to bring the sniper rifle. I could pick off a few Japs at long ranges, 600–800 yds, across a valley. I would not stay any more than half an hour popping away and then get out while I was still lucky. I was also used a bit when we were watching

> the Japs in a village ... they would turn me loose. After a dozen shots you would have them [the enemy] running around like ants.

Late in 1943 in the Finisterre Ranges in New Guinea, Latimer, learning that there were Japanese soldiers in a camp near a local native village:

> decided to stir them a bit. The Intelligence section had a portable range-finder and the night before an escort took me there and I selected a possie [position] overlooking the village. The range on the range-finder was 800 yards and I got set up and they left me to it. I stayed most of the next day shooting up the village and anything that moved. I know I got 6 definite hits which was my best ever for a day, but I really stirred the Japs. They used a lot of mortar ammunition trying to find me.

The Japanese had experienced jungle fighting in Manchuria in the 1930s and during the Sino–Japanese War that had begun in 1937. They were taught to live frugally, carrying rice and water to ensure self-sufficiency. Consequently, Japanese snipers, selected for their marksmanship, were capable of staying hidden and surviving on very little food for long periods. Becoming a sniper was seen as honourable, men seeking at all costs to avoid the shame of returning to infantry roles. Such training as was given focused on ambushing, camouflage and field craft, as the snipers were not widely used as scouts and intelligence-gatherers. In Martin Pegler's *Out of Nowhere*, Tomoyuki Sashiro explained:

> Our job was to hold up the American advance, and to kill as many as possible before we too were killed. Dying in the service of the Emperor was an honourable thing but we should not take unnecessary risks for the longer we stopped the enemy, the greater the achievement.

Compared with Japanese snipers, however, Americans and Australians had better equipment, more effective back-up once aerial superiority had been achieved, and far greater training for their role.

Across Europe to the Reich

Lessons learned in the Pacific, North Africa and the Mediterranean ensured that, by D-Day, both the US Army and the Marine Corps had expertise which could be put to good use: not only on the beach-heads, of course, but also as fighting concentrated on the *bocage*.

Sergeant William 'Bill' Jones of the 8th Infantry said that this comprised a 'patchwork quilt of little fields enclosed by hedges of hawthorn, brambles, and vines tangled together to make walls ten to fifteen feet tall. A lot of the hedges grew out of big mounds three to four feet high, with drainage ditches on either side'. Fighting in such an environment proved to be not only difficult but costly. Casualties mounted rapidly, Jones continued:

> The sick-sweet stink of death seeped into everything. If it wasn't dead Germans and sometimes a dead GI, then it was dead cows all bloated up or a horse or goat or something that got caught by artillery or in a crossfire. I must have seen five hundred dead cows.
>
> Sometimes you discovered a fresh shallow grave topped with a wooden cross or a boltless rifle stuck muzzle-down into the mud with a helmet on top of it. I don't know how many folks died in the hedgerow fighting after D-day, but there had to have been a lot of them. We just kept going and kept going. It was the same thing day after day. Mostly we didn't know where we were going and some of us forgot where we started from ... War in the hedgerows came down to just what you could see from one hedge to another.

Jones observed that the *bocage* was:

> tailor-made for ambushes and snipers. The lanes were death traps as much for tanks as for dogfaces. The Germans had machine guns and snipers mounted on wooden platforms in the trees. The snipers used flashless gunpowder. That made it hard to spot them. They kept picking us off. It got to where you suspected danger in every blade of grass.
>
> You dreaded the next bend in the road. It got where you had almost rather be shot on the spot and put out of your misery than have to go around another bend or across another field and face another machine gun or another sniper picking you and your buddies off one by one.

However, the Allies managed to break out of Normandy and begin the long hard slog across northern France by way of locations whose names are near-mythological – Arnhem, Bastogne, Nijmegen – while the Red Army thrust ever further westward. By February 1945, the Russians had crossed the Oder and US units had reached the Siegfried Line. Harry Furness commented that:

> Much of the advance was spent in attempting to deal with German snipers, and while many were very experienced, a large number were not, having been rushed to the line as soon as their training was complete. One German sniper commented that: 'we had only just finished our training when we were sent to Normandy. In the first week four of my comrades were killed and I was wounded by a grenade that was thrown into the roof in which we were concealed. I hid my rifle and surrendered my pistol and so I survived. I was eighteen years of age'.

Snipers on both sides were involved almost on a daily basis, though the Germans, ever more greatly reliant on Volkssturm (the last-ditch volunteer forces formed in October 1944 to defend the Reich, manned by virtually anyone who could stand and fire a gun) and individual volunteers, began to run short of equipment as the Russians retook occupied areas in which the Germans had exploited arms production. A Waffen-SS veteran recalled that:

> Our unit morale was good, we understood the war was lost and most of us had no homes because of the bombing, but we were not going to surrender. The enemy would pay for every metre of ground. In these months I gained very much experience and made many scores but the British snipers were well equipped and hard to fight. Usually, we made a position and fired, maybe one or two shots, then left quickly for it was dangerous to remain. One bullet smashed the sight on my rifle and I could not get another for some time so I was a sniper without a telescope. I was very lucky to return to Germany, but there was nothing left.

On 21 April, Red Army units entered the outlying suburbs of Berlin: the end was in sight. But victory had a terrible price …

WHAT MAKES A SNIPER?

A document published in Germany in 1915, regulating the use of non-military rifles which had been pressed into temporary service, said that the marksman (there is no specific word for 'sniper' in German):

> must shoot with discretion and the rifles must not be fired for salute or suppressive fire. Marksmen are not limited to the location of their unit, and are free to move anywhere they can see a valuable target. Sentries have the duty to signal the marksman, such targets they themselves can determine.
>
> The marksman will use his telescope to watch the enemy front, recording his observations on a note book, as well as his cartridge consumption and probable results of his shots.

Responding in *Red Army Sniper* to the question 'so what is a sniper?', Evgeniy Nikolaev suggested that:

> A sniper is both an accurate sports shooter and a hunter, and also a rare but established element in every military unit during both peacetime and wartime. The modern sniper is a rifleman who is, as distinct from other soldiers, armed with a rifle with telescopic sights. He is a specially trained, tactically aware, particularly accurate (or sharpshooting, as he is often termed) rifleman, who has a superb command not only of the art of sharpshooting but of camouflaging too, plus an excellent knowledge of his weapons components. And he must be not only highly skilled at camouflage, but also capable of detecting through the most insignificant signals an enemy sniper no less well camouflaged than himself. A sniper must know how to observe, know how to draw appropriate conclusions from what he has seen and report it to his commanding officer, to be a real reconnaissance scout.

And Clifford Shore, writing in 1948 in *With British Snipers to the Reich*, offered opinions which would not have seemed so very far out of place in the First World War:

> The general maxim to the sniper was that he should attain the standard of the highly skilled and cunning poacher. In operation role the snipers were sent out to the various Companies before dawn and came back at dusk; they did not usually stay with their duty company but were pulled back so that they were ensured as good a night's sleep as possible. Frequently the Company to which the snipers were sent provided a section to give them covering fire when they went into No Man's Land in order to ensure that they were not surprised.
>
> The same snipers were invariably sent to the same Company since it was apparent that each got used to the ways of the other, and, in addition, the system fostered a competitive spirit which was all to the good of the game.
>
> The snipers' task was invariably the same – to dominate No Man's Land and the enemy's forward defended localities when possible by killing Huns and making sure that the enemy's movements were restricted.

Shore also remarked that sniping was not 'the vague, haphazard shooting of the unknown in a sort of detached combat', but instead the:

> killing of a man in cold blood, and is an art which must be studied, practised and perfected. I often heard it said that a sniper should be a man filled with a deadly hatred of the Hun, or enemy. But I found that the men who had seething hatred in their hearts for all things German, such as those who had lost their wives and children and homes in the blitzed cities, were not the type to make good rifle killers. The type I wanted was the man of cold precision, the peace-time hunter who had no hatred for his quarry but just a great interest in the stalk and the kill!

Body, Mind and Spirit

Analysing the psychology of the sniper is still contentious, as the popular view of snipers as little more than cold, calculating, emotionless killers can be misleading. Undoubtedly, some men were sociopaths and others were 'loners', but many were seeking to honour their cause or country – even though they may have been conscripted. In addition, large numbers were seeking revenge for loss of property or family … and sometimes also to exact the most brutal revenge for atrocities real or perceived.

Physical fitness and high levels of stamina were essential, as snipers were often called upon to endure conditions which would defeat many men. Another problem was reaction to the often arbitrary matters of life and death. Ion Idriess, a first-class shot who was to become not only Billy Sing's observer but also an effectual sniper in his own right, reflected after a man had been killed in front of him:

> He was a little infantry lad, quite a boy, with snowy hair that looked comical above his clean white singlet. I was going for water. He stepped out of a dugout and walked down the path ahead, whistling. I was puffing the old pipe, while carrying a dozen water bottles. Just as we were crossing Shrapnel Gully he suddenly flung up his water bottles, wheeled around, and stared for one startled second, even as he crumpled to my feet. In seconds his hair was scarlet, his clean white singlet all crimson.

And Frederick Crum dwelled at length in *Memoirs of a Rifleman Scout* on the demise of Second Lieutenant William Ryder Bird:

> Our snipers have been doing good work, but, alas, we lost a fine young officer, shot through the head by a sniper. It brought it very near, and somehow, though so near, it seemed to me nothing but a good end as far as he was concerned, keenly doing his best, knocked clean out, no pain, straight to the not very far beyond where men go who do their duty. It was his very keenness that ended his time with us.
>
> He had asked me to help to put in loopholes in a rotten parapet he was rebuilding that night, not 30 yards from the Hun crater. I had been looking round and talking over things with him; we had cautioned him about exposing himself too much.
>
> I was looking over the parapet with my periscope and talking to him; he jumped up and, without my noticing, peeped over just on my left. There was a crash and a smack, and I realised he had been hit. The bullet passed right through a rotten sandbag in the top layer and got him low down in the head. He died as he fell, without a groan or a word. The bullet passed on through a subaltern's cap and splinters of it hit my good Corpl. Scott in the face.

Crum then reflected on another incident in Blangy, 'at a point where the Main Street ran through the village':

> It gave me a bit of a turn to see the silver outline and black centre of a rifle barrel pointing in my direction. It seemed so close. I kept quite still. Then, slowly, to my relief, it moved away from me …
>
> I then went back to warn the Sergeant on duty that he was to caution everyone who passed to keep down, and get the trench deepened that night. As I went, a shot rang out, and I turned to see both my Corporals wounded in the bottom of the trench. Otterwell died in my arms almost immediately … Riches was not severely wounded. From that time onwards a special vendetta existed between the opposing snipers …
>
> My dear Corporal Otterwell … is such a loss I feel it every day. A fine young fellow, just going home for his Commission. I held his head up as he passed away, and realised as I saw him going what a useful young life was passing – no pain, a soldier's end, a gallant lead and duty done to the last. I felt no sort of feeling of revenge, and yet I tell my snipers it is their duty to do all they can to hit back …
>
> After long planning and watching, Otterwell was at last avenged by our left-handed champion, Sheehan, who killed his man and silenced all sniping from that quarter.

Klavdiya Panteleyeva reflected that she and her partner Marusiya had:

> placed our rifles at one embrasure and watched the German defenses from the breastwork. But the Germans also put a sniper to watch us. And so I was watching, observing during my shift (because the eyes would get tired), and Marusiya said: 'Let me take the watch now'. She got up, it was a sunny day, and she apparently moved the lens.
>
> As soon as she got up, there was a shot, and she fell. Oh, how I cried! The German was 200 meters away from us. I screamed so loud it could be heard all over the trenches, soldiers ran out: 'Quiet, quiet, or they'll open mortar fire!' But how could I be quiet? She was my best friend.

Experiencing the death of colleagues was bad, but even killing an opponent wasn't always straightforward. Lieutenant John George of the USMC noted that his first shot in combat:

> went too high, at least a foot over his back according to Art [his observer]. The Jap fell flat on his belly the instant my shot went past. Had Art not

> been there with the big telescope, I would have given myself credit for a hit, but he saw the bullet hit the sand in the wrong place. It was luckily one of those rare instances where a sniper is afforded a second shot after he has muffed the first one. The Jap remained motionless on his belly, possibly thinking he was being fired on from our lines below but from my position above he remained an excellent target I fired the four remaining shots in the rifle, getting at least two hits.

However, kills came easier to some snipers than others. US Army Sergeant John Fulcher, on at least one occasion, drew on his part-Cherokee heritage to scalp a German corpse in a successful attempt to demoralise his opponents. He noted that even before leaving the USA for Europe, landing at Salerno in September 1943, he knew he had the skills:

> to kill a man up to five hundred yards away. What none of us knew, however, was how we'd react when we peered through our scopes for the first time at the face of a man we were going to kill with a press of the finger. You can shoot at bull's-eyes and man outlines all you want, and you can bullshit and act bold and boast about what you're going to do, but you'll never know what you've got inside you until the real thing comes. You'll never know if you can do it or not until then.

And John George reflected:

> I nudged his chin with the broad [crosshair] post – a measure calculated to plant the bullet somewhere on his chest at the 350 yards range. Then I gave the last pound or so of the three pound trigger pull a 'gentle snatch'. The scope settled back in time for me to see the bullet strike the Jap and splash sand behind him. I cannot recall the least bit of thinking I had just killed a man for the first time. All I remember is a feeling of intense excitement – the same as that experienced when one downs a hard sought big game trophy.

Yet even the most battle-hardened snipers could be affected by individual incidents, particularly if these had happened early in their careers. One example was Gebirgs-Jäger sniper Sepp Allerberger, who hated killing horses on the battlefield:

> And those who had witnessed his first shot as a marksman were curious to see its results, and they ran over to the woodpile behind

> which their opponent had been lying. He had dug a trench-like hollow in which his lifeless body was now lying. Only the feet were visible. Next to them there was a seeping trickle of blood. Grabbing his ankles two riflemen pulled him out so that the extent of his deadly wound could be seen. A bloody mess of brain and bone splinters was covering his back. In the back of his head there was a gaping hole as big as a fist so that they could see the inside of his skull, which had been cleared out by the shockwave of the bullet. The Landsers, who were accustomed to such sights, turned him on his back to reveal the face of a boy, who may have been 16 years old … Sepp looked down at his victim with a mixture of pride, horror, and guilt. Suddenly he felt a lump coming up his throat and he had to puke.

Even the most experienced sniper instructors sometimes struggled to explain motivation, though virtually all agreed that a desire for revenge, unless carefully channelled, almost always led to poor practice. Clifford Shore commented in *With British Snipers to the Reich* that:

> Personal feelings of remorse or questioning of motives will slow down a man's critical killing instinct and the sniper who allows himself to fall into such a train of thought will not last long. It is imperative to look upon the killing of an enemy as swatting a fly, an unthinking, automatic action.
>
> Two things only should really interest the sniper – getting the job done and getting away unscathed. To become accustomed to 'sniper killing' is not so difficult or hard as close quarter killing. A man dies more slowly than the average person thinks; he often grins foolishly when he's hit, the whites of his eyes roll upwards, death sweat gleams on his forehead and he sags to the ground with a retching gurgle in the throat – and it is difficult to hear that gurgle without emotion.
>
> The sniper is usually spared all this unless he's quick enough to get his binoculars on his victim, or the 'sniped' is at such close range that the telescope sight will give him all the motion pictures he needs.

But he also acknowledged that there was a huge gulf between simply talking of killing someone and the act of taking the shot.

> When one was in position there came an 'inner freezing'; the breathing was not quite normal; the hearing sense was magnified and there came too that sense of excitement which all hunters know and which results

> in an unconscious nerve-hardening, and once the Hun was in the sight and the pointer steady at the killing spot there was no qualm of conscience about hitting him or taking life.
>
> The true hunter is never a butcher; he does not desire to kill for killing's sake, but there is something elemental in the stalk and the slaying which swamps every other feeling and makes the heart and brain exultant, and filled with action-elation.
>
> It always seemed an unconsciously long time from the impersonal squeeze of the trigger to the shot going home and the Hun sagging slowly to the ground. Then followed two emotions. First, the heart leapt; and then one was very quiet inside.

And then there was public perception to consider:

> There are still many unenlightened people who think that sniping is 'dirty', 'horrible', 'unfair'. I maintain that it is the highest, cleanest game in war. It is personal and individualistic; a game of great skill and courage, of patience and of forbearance. And if one loses to a skilful opponent one should never be conscious of it. Death-night to the sniper's quarry should always be of a tropical, sudden intensity.

Sniping wasn't always easy, owing to the need to keep alert for long periods of time – often in the most inclement weather – and to avoid making mistakes resulting from fatigue, hunger and many other factors. Frederick Crum recalled in *Memoirs of a Rifleman Scout*:

> Today it has poured and poured, and the mud which was 'the limit' before must now be beyond all words The guns keep going through it all, and I suppose the troops of both sides are being put to as a high a test of endurance as ever is reached in war. My own occupation of late has been centred in supervising sniping and with visits to the trenches at night.
>
> You can hardly imagine the mud. One squelches and struggles along in gum-boots, slipping, slithering and stumbling in the dark, saving oneself from a fall at the expense of covering one's hands inch-deep in mud, or, failing that, taking a pearler [sic] and covering oneself and one's clothes from top to toes with liquid filth, which one carries back to the dug-out, impossible to clean up self or clothes.
>
> I think I mind the mud far more than any shells or bullets, though I do not pretend to enjoy them either.

And recruitment was sometimes a matter of chance. As the Nikolaevs state, Ilya Grigorev did not immediately find his vocation, but, after picking up a sniper's rifle one day:

> went into a hideout with a young partner. In the trench, which they had reached during the dark, Grigorev pulled some empty food tins out of his pockets and hung them on a wooden hoop. 'What does he need the hoop for?', the young sniper wondered.
>
> And then he realised: his senior partner had thought up yet one more ruse – he was a master at this. When daybreak came and the mist cleared over the enemy trench, the snipers saw a Nazi leaning over the parapet.
>
> 'I'll get him', said the young sniper, giving Grigorev a questioning look. 'Hang on', answered Grigorev grumpily. 'There's something not quite right' … The Nazi hung around for a minute and disappeared. However, he reappeared after a short time, but in a different place.
>
> By means of some special sense Grigorev had guessed the enemy ploy: they were facing a dummy.
>
> 'That won't catch us out, old boy', said Grigorev with a smile and turned to the young sniper, 'Go further along the trench and bowl the hoop without being noticed, so that it rolls. Understand?'
>
> 'I understand everything now, Comrade Starshina [sergeant major]!'
>
> Soon the food tins could be heard clanking in the morning silence. And, naturally, the Nazi sniper was seized by curiosity … He cautiously looked out from his refuge. This was enough for Grigorev to momentarily scan the Nazi's face in the crosshairs of his sights and press the trigger.

One of the most important requirements in a sniper is good eyesight. Eyes are rarely the same, but the differences in an individual can be great enough to affect binocular vision and, therefore, the ability to gauge range effectively. This is easily tested, but peripheral vision, hand–eye co-ordination, reaction time and innate spatial ability may not be so obvious. Some men with excellent vision proved to be poor performers when speed of decision was involved: shooting at disappearing targets, for example, or wing-shooting. However, the best snipers have almost always had extraordinary abilities.

The 'normal' eye can resolve images occupying a minute-of-angle (1 MOA) at a distance of a hundred yards, but tests undertaken on more than a hundred Soviet marksmen in the mid-1960s showed their eyesight to be significantly better. Marksmen who had been shooting for only two years recorded 0.704 MOA while those with ten years of experience recorded 0.602.

Better-than-average visual acuity explains why some snipers have always had a reputation for seeing things that their colleagues did not. Not for nothing were the Red Army's Nikolay Ilin and Gennadiy Velichko nicknamed 'Hawkeye' and 'Golden Eye' respectively.

In addition, visual acuity could be improved by practice even if the firer's vision was judged to be above average. Tests also showed that visual acuity recovered better in some people than others while resting after intensive practice. Rapid fire was found to reduce visual acuity sharply in many people, and maintaining aim for long periods without blinking or shifting focus could ruin accuracy. Martin Pegler remembers how his instructors advised that looking at trees, bushes or grass every ten minutes could benefit the sniper, as brown and green colours were known to relax eye muscles.

Physical fitness was another important asset, especially when the sniper's task involved periods of inactivity while observing his target. Consequently, most sniper-training courses included self-defence classes, football and similar sports to ensure that their trainees reached – and stayed – in peak condition.

The ability to sustain minor wounds without affecting performance was also valuable. Snipers have always sustained heavy casualties, yet many individuals managed to shrug off injuries from bullets and shell fragments. Red Army sniper Bella Morozova, for example, survived even though a bullet which hit her in the temple penetrated the nasal cavity and destroyed one of her eyes.

However, many snipers were eventually reduced to training roles when the long-term effects of the wounds reduced their effectiveness: Russians Iosif Pilyushin, Ivan Sidorenko, Evgeniy Nikolaev, Lyudmila Pavlichenko and Vasiliy Zaytsev among them.

Calmness under pressure was yet another useful trait, allowing the sniper's breathing to be regulated and heartbeat to be slowed. Shooting with greatest precision could not be done if natural tremors affected aim. However, instructors often found that many otherwise excellent candidates panicked when required to take a difficult shot, or alternatively were so excited that their innate natural caution was overriden in the heat of the moment.

Something which could rarely be predicted was the reaction of a sniper when faced with his first human target. For many, the natural reaction was to hold fire. Looking into a man's eyes, knowing your shot will kill him, does not always come easily. Countless men surrendered their lives simply because they 'froze' just long enough for the intended victim to shoot first. Even highly experienced snipers – Henry Norwest, for one – were killed because their shot was a fraction of a second too late.

Norwest's shot, fired while the fatal bullet was in flight, is said to have killed his German opponent. There have been other well-documented mutually fatal encounters, and still more where a fatal result was avoided only by good fortune such as the bullet striking the rifle or its sight.

It is important, too, that sniper and observer are both attuned to the conditions; someone brought up in the snows of the Arctic will not necessarily cope with the heat of the desert. Australians fighting in Gallipoli in 1915 coped well enough with the daytime heat, but the cold nights gave rise to hypothermia and even mild frostbite.

There was no point in expecting a man from the northern wastes to perform as well in the tropics without lengthy acclimatisation. Clifford Shore offered an explanation for why 'the standard of shooting in the hot climates was found to be lower than in the temperate fighting zones':

> It was impossible to produce a high standard of shooting when the men had to shoot in shirt sleeves, the causes being – one's elbows must have a comfortable position on the ground; with only a shirt covering the elbows when crawling and shooting on hard and stony ground life, for the sniper, was not much fun; the rifle must have a firm bedding in the shoulder, and this can never be obtained with any rifle when there is only the thickness of a shirt between a steel rifle butt and the shoulder bone … with a sniper's rifle it is even more difficult since due to the height of the 'scope sight the firer's head is forced into a position high on the cheek rest and consequently the end of the butt is forced down into the shoulder that amount; this makes it difficult to bed the rifle securely in the shoulder even with a sufficiency of clothing.

Snipers were also often required to maintain a lengthy vigil with very little sustenance. Simo Häyhä sucked sugar cubes to maintain his metabolism, while Mikhail Kondratyuk – detailed in Evgeniy Nikolaev and Dina Nikolaeva's book *Sovetskie Snayperu v Boyo 1941–1945* – ate as many carrots as he could find in the belief that such a diet sharpened his eyesight. Dehydration could present serious problems, even in temperate zones, and a well-filled water bottle was essential to any lengthy foray.

RECRUITING THE SNIPER

In the early stages of the First World War, particularly on the British side where the concept of sniping was seen as a waste of time at best or 'ungentlemanly' at worst, the amateurs held sway. Most were officers with experience of target or game shooting, until the terrible lessons were learned.

Efforts were then made, unofficially at first, to recruit fieldsmen – Scottish ghillies, beaters and even poachers; Canadian Native American trackers, hunters and guides; Australian kangaroo hunters; big-game chasers from India, southern Africa and the British aristocracy.

Men of the Lovat Scouts, drawn from northern Scottish 'great estates', had proved their value in South Africa and were to be of great service in Gallipoli and on the Western Front in the First World War, and then again during the Second World War.

Some men saw sniping as a way of settling scores, often avenging the loss of family members, friends and colleagues, or for atrocities real or perceived.

There were those whose motivation stemmed largely from duty or patriotism, and not a few whose rural background – sometimes amplified by racial discrimination – suited them to the chase and the kill. Frederick Crum reflected in *Memoirs of a Rifleman Scout* on the quality of his trainees:

> What men they were! Some two dozen picked Riflemen. The names that occur to me are: Scott, Evans, Otterwell, Liddiard, Sherry, Hicks, Riches, and others; Corporals then, but later mostly commissioned officers, and today – where are they? Every one of them killed, wounded, or missing!

But Eric Penberthy had sounded a cautionary note in 1920 in *The English Review*:

> It is interesting to note that experience showed us that a good 'competition' shot of, say, a high Bisley Meeting standard did not

> necessarily make a good sniper. There is the width of the world between the leisurely peace time business of firing at a distinct target at a known or easily calculated range, after studying wind, temperature, mirage, light, the error of the day, and all the mysteries which make the Bisley Meetings such a pure delight to the 'pot hunter', and getting one quick and accurate shot at such almost invisible and momentary targets as the Boches presented in trench warfare. The fact also that in many cases the Boche was shooting too made it a short, sharp duel, the winner being he who got his shot in first, and the loser probably losing his life as well as the trick. The realisation of this called for nerve and courage of a high order. All the same, old snipers will tell you that it was this very risk which gave 'the game', as they called it, its great fascination.

The desperate need of men to counter the depredations of Germans on the Western Front and the Turks in Gallipoli initially led to encouragement of volunteer snipers within the rank-and-file – often experienced shots serving away from the front line. They included Sergeant Thomas Brennan, in charge of the cooks of the 7th Light Horse AIF:

> Every morning, breakfast over, he took down his rifle, strolled across to the trenches, and had innumerable duels with Turkish snipers.
>
> He had the range of all their trenches and when he saw a sniper's 'posey' he blazed away till he silenced the enemy. Now and then an unwary Turk showed half a head, and this Irish sharpshooter was on to him like a shot. Some days he would come back to camp angry and disappointed.
>
> 'Thirty shots and not a single scalp', he exclaimed, kicking aside some innocent mess tin.
>
> But at other times he stalked back as if he had won the battle of Anzac on his own. 'Killed three Turkeys', he cried. And then he was happy as Larry all day.

And Eric Penberthy observed in *The English Review* that:

> On one occasion several runners had been sniped, and at last Major Garnet Hughes (now General) went round to the cook-house and called the cook, a man noted for his marksmanship and a winner of the D.C.M. in South Africa. He said, 'Take your rifle and shut up that sniper'. The cook dropped his spoon and was away some twenty

> minutes. He came back with a grim smile, and the Hun sniper was silent for ever.

Another volunteer was the Reverend Andrew Gillison, who, a few months before war began, had contested the King's Prize shoot-off at the Williamstown rifle range in Melbourne. Gillison obtained the best score from ten shots at 800 yards, but a cross-wind on the 900-yard range had lost him the prize. Captain Gillison then joined the 14th Battalion of the AIF, sailing for Gallipoli to follow his calling as a chaplain until dying in hospital in Lemnos from wounds. An obituary published in *The Argus* on 2 October 1915 reported that:

> Every day he went along the firing line, telling stories and smiling, until others laughed outright. Now and then he took a rifle into one of the sniper's posts – the same sort of weapon as that with which he had fought for a place in the 'King's 100'.
>
> He did not fire many shots, but every time he pulled the trigger the bullet scored a bullseye.
>
> 'I must have killed nine today – poor fellows', he would say. 'I hope none of them had wives and little ones'.
>
> He had a soft spot in his heart, even for the enemy.

Very different problems were posed by the discrimination and mistreatment that mixed-race snipers all too often encountered in civilian life. Australians Harry Freame, Billy Sing and Caleb Shang all had Chinese fathers, while Joseph 'Paddy' Riel of the 18th Battalion of the Canadian Expeditionary Force (CEF) (killed in January 1916 with 30 kills to his name) was said to be 'Métis' – a mingling of the blood of the earliest European settlers and Native Americans – though of largely French and Irish ancestry.

Henry Louis Norwest ('North West' or 'North West Wind'), part French-Canadian, part Cree, enlisted as 'Henry Louie' to serve with the 3rd Canadian Mounted Rifles. Discharged for drunkenness, he spent time in the Royal Canadian Mounted Police before re-enlisting in September 1915. The war diary of the 50th Canadian Battalion duly recorded that, on 29 April 1918, 'Sniper H. Norwest M.M.':

> today accounted for his hundredth German, and before night had added another to the score, a record enjoyed by few, if any, other snipers in the British Army. Norwest's methods are peculiarly his own. 'Wait until not a single Hun has a chance of seeing your rifle flash, then get your man', is his motto.

Credited with 115 kills, Henry Norwest survived unscathed until killed while counter-sniping in August 1918.

Many of the Canadians from the trackless expanses of plain and forest, who took a dreadful toll of Germans from 1915 onwards, were pure-blood Native Americans. Eric Penberthy noted that:

> One of the finest snipers I ever met was a full-blooded Red Indian – John Ballantyne [sic]. He applied all the methods of the chase, so familiar to him in his beloved Canadian forests, to hunting the Boche sniper. He had been known to wait patiently for seven days in a wonderfully concealed and prepared sniper's post for a valuable target – a Hun officer, whom he finally killed.

John Ballandine of Battleford, Saskatchewan, was widely regarded as the 'champion Canadian sniper', serving the 32nd Battalion CEF until seriously injured in April 1916 and subsequently discharged. His brother James also served as a sniper, alongside Akwesasne Mohawk Philip McDonald of the 8th Battalion CEF (killed in 1916 with 70 kills to his name) and Johnson Paudash (88 kills) of the 21st Battalion CEF, whom the Canadian census of 1911 described as Chippawa but is now generally considered to be Ojibwa (Anishnaabe). Herbert McBride, in *A Rifleman Went to War*, said of Paudash:

> By birth, inheritance and inclination, he was a hunter. He never would have an observer, preferring to work alone as he made his devious way along behind the lines, waiting for a chance to take a shot. He seldom (perhaps never) had any permanent 'nests' but moved about continually. Each evening he would turn in his report, and I for one believed him, which is more than I could say for any other lone sniper.

CEF sniper Francis Pegahmagabow, part Ojibwa, specialised in night raids. His score mounting rapidly, 'Peggy' was awarded the Military Medal in 1916, followed by Bars granted for bravery shown during the Battle of Passchendaele in November 1917 and the Battle of the Scarpe in 1918. A tally of 378 'confirmed kills', if true, makes him the most successful sniper of the British and Empire forces.

However, many factors had to be taken into account when considering recruitment of ethnic groups, some of which proved to be better suited to service than others. Hesketh-Prichard, commanding the British First Army's School of Scouting, Observation and Sniping, observed in *Sniping in France* that:

> The Germans, with certain brilliant exceptions, … were quite sound, but rather unenterprising … The Bavarians were better than the Prussians, while some Saxon units were really first rate … The English were sound, exceedingly unimaginative, and very apt to take the most foolish and useless risks. The Welsh were very good indeed. The Americans were also fine shots, and they thoroughly enjoyed their work. The Canadians, the Anzac and the Scottish regiments were all splendid, many units showing an aggressiveness which had the greatest effect on the morale of the enemy.

He also noted that pre-1918 British Army snipers held an advantage over their opponents, even though:

> In the very crack Jäger regiments, such … as were, I suppose, recruited from Rominten or Hubertusstock districts, where the great [hunting] preserves of the Kaiser lay, and in which were a large percentage of Forest Guards, this [shooting skill] was very noticeable. But for long-distance work, and the higher art of observation, the Germans had nothing to touch our Lovat Scouts. This is natural enough when one comes to consider the dark forests in which the German Forest Guards live … Mostly game is seen within fifty or seventy yards, or even closer, in these sombre shades, and then it is only the twitching of an ear or the movement of an antler lifted in the gloaming. Compare the open Scottish hills. It was the telescope against the field-glass, and the telescope won every time.

Another obvious source of recruits were youth organisations such as the Boy Scouts and the Boys' Brigade. But though the Boy Scouts had included basic marksmanship training in their curriculum when founded in 1908, and though the Hitlerjugend followed suit during the Third Reich, only the Soviet Union promoted state-sponsored marksmanship in the inter-war period.

In 1931, the Central Committee of the Communist Party ordered that basic military training should be mandatory for all boys and girls who entered elementary school. Little Oktobrists, aged 7–9, became Young Pioneers (10–15), who in turn could join the ranks of the All-Union Leninist Young Communist League – Komsomol – which had been founded in 1918.

The Union of Societies of Assistance to the Defence and Aviation-Chemical Construction of the USSR, 'Osoaviakhim', had been created on 27 January 1927. Its remit included training recruits for the armed forces: by 1928, 5,297 'Shooting Circles' had 230,000 participants. Introduced on

29 October 1932 to replace a variety of old badges and awards, Osoaviakhim's Voroshilov Badge had a junior and two senior stages. More than 36,000 graduated in the first year; and, at the 1936 Voroshilov Rally, another 215,000 trainees were recruited.

The marksmanship badge often proved crucial when individuals were either rejected for military service or training as snipers. Aleksandr Romanenko was among them:

> I immediately asked to join the snipers but, because of my short stature (I hadn't fully grown by then), they did not want to enrol me in a cadet sniper company. It was then I was compelled to produce my 'Voroshilov Marksman' badge and accompanying documents. After that they didn't have the right to refuse me.

Formed in 1918, Vsevobuch (the Universal Military Training programme) was abandoned in 1923. In June 1941, however, the programme was reactivated to provide a minimum of 110 hours of marksmanship and similar instruction for civilians over the age of 16. Almost ten million people graduated from Vsevobuch courses between 1 October 1941 and 8 May 1945.

Many of the most successful Russian snipers were ethnic-minority hunters acclimatised to the harshest of conditions. They included Maksim Passar (1923–43), from Nizhny Katar in Siberia, who is customarily overlooked for his valuable contribution to the defence of Stalingrad, where most of his 237 kills were achieved. Mortally wounded in January 1943, Passar had trained many snipers for the 117th Rifle Regiment who between them amassed 775 kills.

A member of the Evenki people, Semyon Nomokonov (1900–73) was born in what is now the Zabaykalsky autonomous region of Russia. He learned to shoot at the age of seven and spent his formative years hunting in the Taiga. When war began, Nomokonov, who had trained as a carpenter, was conscripted to make crutches for the wounded. However, no sooner had Nomokonov begun work than he quickly demonstrated his prowess with a rifle, by shooting first and so getting the better of a German soldier who had been aiming at him.

A rapid transfer to a sniper unit was arranged, followed by service as a sniper and sniper instructor in places as diverse as Karelia, the Ukraine, East Prussia and Manchuria. His logbook is claimed to register 360 kills on the Eastern Front and seven or eight Japanese soldiers killed on the Transbaikal Front in Manchuria after the end of the war in Europe.

Some snipers, often highly successful, began to ply their trade only after the war had begun and the value of marksmen could be seen at first hand.

As detailed by the Nikolaevs, Nikolay Ilin, for example, only became a sniper because he:

> served in the same unit as sniper Vasily Silin. A good bloke, no denying it, and a good shot. I asked Silin for help. Well, understandably, he told me all the finer points and then I studied the properties of a sniper's rifle and telescopic sights. I made myself familiar with the laws of ballistics, practiced as required, developed the knack, and gradually things started to fall into place. And it was then I realised: you won't get to be a sniper without effort and sweat. And now I'm no longer on my own; there's a whole platoon of us. All with the right fighting spirit, as if they'd been specially picked. Take Lapa, for instance. Here at Belgorod he's already wiped out eighty six Nazis, and Gordienko's got seventy five. It all adds up to a tidy sum. We are fighting for the whole battalion.

The best known of the male snipers is undoubtedly Vasiliy Zaytsev, author of *Notes of a Russian Sniper*, reprinted many times, and a central figure of the book *Enemy at the Gates*.

When war began, Zaytsev, a petty officer clerk in the navy stationed in Vladivostok, volunteered to serve in the front line. Transferring to the army as a senior warrant officer, after proving his capabilities as a marksman – skills learned at an early age in the Ural mountains – Zaytsev was posted to the 1047th Rifle Regiment, which on 17 September 1942 was attached to the 62nd Army ordered to defend Stalingrad.

Fighting among the shattered ruins of a large and industrialised city, making legends of sites such as the Red October factory, not only presented the snipers with unique challenges but also tested their ingenuity to its limits. Zaytsev, attempting to neutralise a German sniper hiding under a wrecked train, reflected on the use of one seemingly inappropriate location:

> I figured he was breathing easy now, thinking that he had killed one of our snipers near the loudspeaker. Meanwhile Gorozhaev and I were crawling to the side, looking for the perfect spot. We were careful to stay hidden from sight. About thirty metres east of the boiler room stood a huge vat that had once held tar, and on top of it sat a platform for a hydraulic hoist. A short ladder led up to the platform. We climbed up and then lowered ourselves down into the vat. The dry tar that still lined the walls of the vat was sticky, and its stench was irritating.
>
> We pulled a board off the platform above and placed it beneath our feet. The sides of the vat had several small holes, and we selected

> the least conspicuous ones for our peepholes. From this location, the railway car where the Fritz sniper had positioned himself was in full view. At that moment, the train wheels obstructed our view of the sniper, but we were willing to wait. Other than the stench from the tar, we had found the perfect position.

Many women volunteered for service immediately the Germans invaded the USSR, sometimes because of patriotism but also to avenge personal tragedies, massacres and atrocities. For example, Mariya Polivanova, whose family lived in Moscow, had lost two of her brothers – killed during the German advance – and a third lay seriously wounded in hospital. She wrote:

> It's a very difficult time at the moment, but we need to hold out. The enemy think we don't have the heart for it. But we do! It's the choice season for snipers at the front. Natasha [a popular 'pet name' for Natalya] and I go off stalking in the forest for Nazi 'cuckoos'. We track them down and wipe them out. Natasha and I have had to face duels with enemy snipers. All defenders of the homeland have one thought in mind: to cleanse the Soviet land of Nazi scum. I shall strive to take revenge on the enemy for the grief and suffering of our people. I do not begrudge my own life in that cause!

There was irony in the closing sentence, as Polivanova, with her partner Natalya Kovshova, was to die in battle.

Attempts to exploit pre-war marksmanship qualifications to enter service were generally thwarted by *voenkomaty*, military commissariats functioning as draft boards, which were manned by old soldiers who rarely saw use for women in the front line other than to become medical orderlies, drive supply trucks or type letters. In addition, volunteering often fell foul of the absence of national registers listing women eligible for military service – a respect in which Komsomol records proved to be of greater use.

Lyudmila Pavlichenko was just one who encountered rejection. Nina Petrova was another; and many other markswomen found their way blocked for a variety of reasons. However, they were to make huge contributions to the defeat of Hitler's Germany: Pavlichenko, with 309 accredited kills, ranks among the most effective snipers of all time.

By losing their lives in battle or due to its consequences, Tatyana Baramzina, Natalya Kovshova, Aliya Moldagulova, Mariya Polivanova and Roza Shanina have all attracted attention. But Nina Lobkovskaya rose to command a sniper company; and Lyubov Makarova shot left-handed very successfully.

Some of the snipers were very young. For example, Klavdiya Panteleyeva was born in 1926. As a member of Komsomol, she was required to attend secondary-education classes which prepared youngsters for military service. Learning that a sniper school had opened, Panteleyeva volunteered in June 1943. She recollected:

> Everyone was 18, and I was 17. They were thinking, should they turn me away or not? Decided that if I didn't fall behind, they would leave me at the school. We started building a firing range. I wasn't from a rich family. I had chopped wood, and carried water, I was accustomed to such business, so I worked well. They let me stay, even gave me leave to visit home.
>
> Then they started teaching us to shoot, but I couldn't do it. I would fire, and all I hit was 'milk' [jargon denoting a complete miss]. Then Zinaida Andreevna Urantseva, our squad commander, started practicing with me individually. She taught me to shoot well. I graduated from the school, and those who graduated with good marks received American presents. My sniping partner was Marusiya Chikhvintseva, from Izhevsk, Udmurtia. We were friends. And so they sent us, many girls, to the front on 1 March 1944.

Aleksandra Medvedeva said that:

> Having learned of the existence of sniping courses, I appealed to my district military commissariat with a request. I was directed to the Central female school of sniper training. There was a whole company of small – both in stature and age – girls. We were even nicknamed like that – 'pencil company'. I successfully finished this school and in the rank of senior sergeant arrived to the front near Orsha in the end of 1943 – to 508th rifle regiment of 174th rifle division.

Antonina Kotlyarova, born in 1923 in Moscow, was studying at the 1st Lenin District School when war began. Her husband's father started organizing militia, while her husband Nikolay joined the armoured forces as a submachine-gunner. Antonina herself took up employment with a machine-tool manufactory. When the factory was evacuated to Nizhniy Tagil, she went to the military commissariat only to be turned away on the grounds that women couldn't fight. After joining Komsomol and moving briefly to Ryazan Oblast, Kotlyarova joined the 50th Anti-Aircraft Artillery Regiment before discovering that there was a sniper school.

Volunteering once again for service, she was immediately accepted; she graduated with excellent grades and asked to be sent to the first Belorussian Front to be near her husband and father. With her partner Olga Vazhenina, and ten sniper-school graduates, Antonina reached the front in November 1944 to serve with 47th Army, 143rd Division, deployed in Poland.

But each story was different. Nina Lobkovskaya, born in Siberia, the eldest of five children of a mining engineer, related how:

> We were all outraged at the news of the war. Young people were very patriotic and longed to defend their Motherland against the invaders, to drive the enemy out. We were bursting to go into action.
>
> At first we only learned of the war from radio and newspaper reports, and we thought that the enemy would be kicked out in less than a year. But the Nazis had advanced as far as Moscow outskirts before they were stopped. In the autumn of 1941 our city then began to receive the evacuated, and soon after echelons with the wounded started to arrive. One by one people left for the front, and they were often somebody I knew or my friends' relations. My father left for the front in 1942. So our family was in it too.

Machine-gunner Aleksey Lobkovsky was then killed in action at Voronezh, causing his daughter Nina to reflect:

> Hatred for the enemy that was causing so much human suffering … cemented my resolution to join the army. But it wasn't that easy – women were not encouraged to fight.

The widespread use of female snipers was unique to the USSR. Though the state's constitution conferred equality regardless of gender, this was widely ignored in the male-dominated army. Consequently, owing to an absence of comprehensive records, exactly how many women volunteered or were conscripted into wartime service is still contested.

In estimates published in 2001, Grigory Krivosheev calculated that 570,000 had been 'called up', 463,503 of whom were active on 1 January 1945. As many of those who had volunteered or fought with the partisans were excluded, Vera Murmantseva suggested a possible total of 800,000, though adding a rider that only about 500,000 may have served in the front line.

Allied Recruitment

As far as the British Army was concerned, lessons learned during the First World War had been consigned to history in 1918 – together with, in many cases, the specialist equipment that was essential to the sniper's job. Writing in *With British Snipers to the Reich*, Clifford Shore reflected on the ineffectiveness of the first steps that had been taken to address recruitment problems:

> When plaintive cries for snipers came from France in the initial months of the war, a sniping school was started at Bisley in early 1940; the War Office's appeal to Bisley for help met with an immediate response, and for a time things went well, and the large majority of the students were jolly good fellows and as keen as mustard. Some of them, however, could not shoot and the riot act had to be read to commanding officers and a threat made that everyone sent to the school would have a shooting test before each course and anyone who could not make a three inch group at 100 yards with the sniper rifles would be sent back to their units.
>
> Then came Dunkirk, heavy air bombardment of this country, and with the Hythe School being bombed out and removed to Bisley there was a certain amount of friction, and eventually the sniping school closed down … And for a time sniping sank back once again into oblivion.

Similarly, the belated entry of the USA into the war in December 1941 found the army and the Marine Corps totally unprepared for combat. Consequently, Japanese snipers took such a terrible toll that the US military authorities were soon forced into taking remedial action. However, this was done largely at local level owing to the absence of standardised guidelines.

As in the First World War, the shortage of trained men was to a large extent made good by impressing those who had already demonstrated their skills in another context. As Charles Sasser and Craig Roberts remarked in *One Shot – One Kill*:

> The best of these marksmen … came from the American Southwest where lanky farm boys grew up as wild as the jack rabbits and deer they pursued year around, or they came from the hill country of Arkansas, Tennessee, Virginia, and Kentucky. Barefooted with floppy hats or wearing dusty runover boots, they were crack shots with their rusty old .22s and .30-30s long before they ever saw a 1903 Springfield or an M-1 Garand.

Gradually, the situation improved as pre-1918 ideas were revisited and the teachings of men such as Hesketh-Prichard and Herbert McBride were dusted off. Though the US Army expanded the training of snipers at a snail's pace, the Marine Corps, with lengthy traditions of marksmanship, saw things differently.

Consequently, scout/sniper training was regularised, when, in the late autumn of 1942, courses began at Camp Lejeune, Camp Pendleton and Green's Farm. These concentrated on honing shooting techniques alongside observational and camouflaging skills, and fieldcraft. Most of the trainees had volunteered, though a few had been sent by their commanders after demonstrating an aptitude for marksmanship.

Though ever-widening recruitment was beneficial, escalating casualties caused problems in all combatant armies. Units were always getting new snipers, men constantly being impressed to make good numbers – but this wasn't always the advantage it may have seemed. Bruno Sutkus noted in *Sniper Ace* that his battalion:

> received a second sniper. It was autumn, it rained and snowed, it was cold and we were soaked through, and we had still not received our winter clothing. We went a whole week without food and lived in our trenches. Several times the Soviets captured our infantry trenches after fierce fighting. We received reinforcements, counterattacked and ejected the enemy. All this caused us great losses, but we succeeded in holding the line.
>
> The new sniper received an order to engage enemy snipers. He found himself a position. His first shot missed, and it was his last because the enemy sniper shot him in the head.
>
> This was a major lesson for me as well: not only my own life, but also the lives of the company to which I had been detached that day depended on my continuing accuracy. Snipers helped out in the most forward sectors of the front and protected the infantry against enemy snipers. For this reason they were feared and hated by the enemy.
>
> Accuracy guaranteed my life. A second newly arrived sniper was seriously wounded a short while later and was invalided out. Thus I remained the battalion's only sniper.

Second Lieutenant Oliver of the Durham Light Infantry poses in 1915 with a Mauser sporting rifle near Houplines on the Western Front, typifying the first phase of countering German snipers. *(Courtesy of Martin Pegler)*

The Warner & Swasey prismatic sight was complicated and somewhat delicate, requiring an experienced sniper to make the best of it. (*Courtesy of Morphy Auctions*)

This British sniper shows clearly the value of camouflage.

Australians at Gallipoli attempt to lure Turkish snipers into firing at a helmet – an old trick which could detect the enemy's position. (*Courtesy of Australian War Memorial*)

The trenches of the Western Front tested the reliability of rifles and sights to their limits, though at least this view lacks the cloying mud of winter.

A German machine-gun post, pictured in 1918. Snipers were encouraged to neutralise these guns by shooting either the crewmen, or, with the aid of armour-piercing ammunition, the gun itself.

Three of the sniper rifles of the First World War: the Austro-Hungarian M95 Mannlicher, the German SG. 98, and the Canadian Ross (*Courtesy of Morphy Auctions*)

Private Thomas Barratt of the South Staffordshire Regiment, the only British sniper to win the Victoria Cross, awarded posthumously.

TRAINING THE SNIPER

Textbooks were written in many languages, but what a sniper needed often came not from any printed page but from personal experience and even instinct. Training had to also ensure that a sniper operated effectively in environments ranging from a devastated city to the vastness of open desert or arctic wastes. However, some men had a 'sixth sense' to warn them of unseen danger while others ventured blindly onwards.

By 1915, it was widely accepted that snipers should be carefully chosen, and also that men who volunteered were more likely to succeed than those who had been forced into the role. Writing in 1941, Nevill Armstrong listed in *Fieldcraft, Sniping and Intelligence* what was needed to make a good sniper:

> An expert rifleman 'out to kill' who is highly trained in observation and the use of ground; equally valuable in trench and open warfare. He should be able to pick out targets exposed for only short periods, and kill with a single shot from concealed positions.

He went on to list several qualifications. The potential sniper should have:

> good eyesight 'and able to use it' and should not be colour blind. He must not be deaf. He should be keen, cool, courageous, persistent, good shot, truthful, patient; and a good scout – with a good eye for ground, particularly in open warfare, and able to read it quickly. Men who were highly strung, of nervous temperament or restless were 'not likely to be good'.

The sniper, Armstrong, added, should have knowledge of:

1. Causes of inaccuracy in shooting and defects in rifle.
2. Use and care of telescope and telescopic sights.
3. Observation.

4. Construction of concealed hides and loopholes.
5. Map reading.
6. Use of compass.
7. Sketching.
8. Value of protective colour.
9. Scouting and use of ground.
10. Night sniping.
11. Penetrations.
12. Wind.
13. Elevation.
14. Judging distance.
15. Messages and reports.

> Selection should be undertaken with great care. Good shooting was essential, but by no means the only qualification. Men must be observant to highest degree, which mean that stalkers, keepers, poachers, prospectors, trappers, out-of-door people, hill and moorland farmers, etc., would be ideal.

Frederick Crum observed in *Memoirs of a Rifleman Scout* that:

> Men selected should be reliable, intelligent, good shots, fit, educated. We are not likely to find the ideal man. The qualifications of the perfect sniper are unlimited. Moreover, the Company Officer naturally does not want to part with all his best men, and there are other equally important branches to be considered such as bombing, signalling or the Lewis guns. On the arrival of drafts the sniping officer should see if there are any likely men … Men who have done well with the snipers have been sent for training at the NCOs school and promoted into various companies. In the past year several of the snipers have been recommended for commissions and most of them have become NCOs.

And Private Albert Huxford of the Royal Berkshire Regiment – then still a teenager – said that he had been selected for sniper training at the Northern Command Sniping School, Rugeley, Staffordshire, because:

> I was always a good shot, shooting was easy for me, but I was never really happy unless I was at a loophole … I was always watching the Germans' lines, and would spend hours in one of our hides waiting for a chance to pot a Hun. I suppose I was a bit of a loner really but I liked

> the element of hunting so when the sergeant told me I was going to be sent on sniper training, I felt it was what I should have been doing all along. I was very pleased.

However, Armstrong also noted that it was useless to try to make a sniper out of every man in a battalion:

> Only a limited number are available, but they must be specialists reserved for sniping. Demands – Patience, perseverance and pluck with precision both of the man and his weapon.

On 11 October 1915, at the headquarters of 11th Infantry Brigade of the 48th Division, using experience gleaned from teaching in the trenches, at Bethune, at Linghem and also from scouting/sniping schools formed by others, Hesketh-Prichard revealed details of the course he had proposed to train snipers effectively. Lasting 17 days, it laid the foundation for much of the sniper training subsequently undertaken throughout the world.

Hesketh-Prichard felt that sniping, in its most basic form, consisted of three main elements: finding your mark, defining your mark and hitting your mark. He also sought to define the role of a sniper, which he felt had never satisfactorily been done before. Many officers believed that snipers were simply to deal with troublesome Germans, but very early on Hesketh-Prichard realised that as much value lay in what the sniper could see as what he could shoot.

In the small training pamphlet written in 1916 for instructors at the First Army School, Hesketh-Prichard defined the sniper's duties:

> 1. To dominate the enemy snipers, thereby saving the lives of British soldiers and causing casualties to the enemy.
> 2. To hit a small mark at an unknown range, but without the advantage of a sighting shot.
> 3. To keep the enemy's line under continual observation and to assist the Intelligence of his unit by accurate and correct reports with map references.
> 4. To build and keep in repair his loopholes and major and minor sniping posts.

The course Hesketh-Prichard founded (details will be found in Appendix 1) and the work of specialist trainers, such as Scotsman Frederick Crum and Danish-born Canadian Peter Anderson, greatly improved the performance of the British and Allied snipers. Most graduates appreciated the effort that had been made to prepare them for combat, Albert Huxford relating how:

> It was refreshing to be treated like a human being, not just a uniform. Many of my fellow students were educated men and I suppose some could have been officers if they had wanted. The instructors treated us with respect and a lot of patience and no one I can recall abused that. The course was hard work but we all enjoyed it and no one wanted to flunk it.

There was also great camaraderie, especially as sniping schools generally saw games and exercise not only as a welcome diversion from lectures but also as a way of achieving the physical fitness a sniper required. The most popular were association football and cricket, often due to the availability of experienced players or coaches in the line, but unarmed combat, which could be the difference between life and death in a trench, was also widely taught.

Nevill Armstrong, in the 1941 edition of his book *Fieldcraft, Sniping and Intelligence*, asked his readers to consider 'first the objects of sniping, then the means by which they may be best attained':

> 1. To protect your own trenches and F.D.Ls., both in positional and open warfare (front, support, communication trenches, and all approaches). This involves –
>
> (a) Obtaining a superiority of fire and keeping down that of the enemy.
>
> (b) Obtaining a mastery over the enemy snipers. Making a thorough visual reconnaissance of the enemy's position and obtaining all useful information.
>
> 2. To kill the enemy sniper and shake the enemy's morale – the two objects go hand-in-hand together.
>
> If the enemy is allowed a quiet time he will utilize it for work of attack and defence. A few well-trained snipers can keep him quiet by their determination with accurate fire and produce an effect that is not measured by the number of rounds fired …
>
> Observation and the ability to describe what he has seen are most important qualifications in a sniper, scout or observer. Good shooting is an essential but by no means the only qualification; men selected must be observant to the highest degree. The sniper should be able to pick out targets that are not obvious to the ordinary view.

He added that:

> Unorganized sniping has proved to be more or less useless. There is no one to guide them, to instruct them, to see that they snipe and not sleep, that they keep the Hun sniper in hand, and that they are taking an intelligent interest in their work.
>
> Battalions who have no snipers or unorganized ones are bound to lose men at the hands of German snipers. In some such instances the casualties were very serious.

And that organized sniping depended on:

> The General Situation. – That is to say, what opportunities are afforded for sniping in the front and support lines or in front of the line and so on. How you will use your snipers. Where you will place them.
>
> Enemy Activity. – Whether he is very aggressive and his sniping good and he has many sniping positions and has good command of our lines.
>
> Nature of Front. – Whether lines are close or far apart. If very close sniperscopes will probably be the only available method, unless we at the same time have sniping positions behind on high ground. If both conditions exist more snipers will be required.

But some commanders saw recommendations for sniper training as a way of ridding their units of mavericks and troublemakers. However, when many men duly returned after failing to make the sniper's grade, their units, and by extension their commanders, gained unwanted reputations of their own. Rapidly, therefore, the best men were sent instead.

Training schools often organised competitions: Hesketh-Prichard, for example, 'had a competition with the prize of a telescope, which was finely won by a small Corporal'. Competition improved not only shooting but also often engendered team spirit, as Private Thomas Durst confirmed:

> Our course had some very good instructors, I recall a Scots Captain who lectured us in map-reading and compasses. I had great trouble with this so in the evenings he coached me and another chap, I recall he often gave us a tot of whisky from his flask to encourage us … in my exam I got nine out of ten on map-reading thanks to his patience.

Becoming a sniper usually allowed an infantryman to have two weeks out of the line. Once qualified, he gained a pay increment and was excused normal trench duties.

Referring to his visits to the British Army's newly established sniping schools, Frederick Crum commented in *Memoirs of a Rifleman Scout* that between 25 May and 2 August 1916 he had been to the First Army Sniping School at Linghem; the Second Army Sniping School; Third Army Sniping School at Albert; Fourth Army Sniping School at Toutencourt and Bouchon; and Fifth Army Sniping School. Typical of events was that held on 2 August at the:

> 4th Army Sniping School, Bouchon: we expect to be here 3 weeks experimenting with a dozen young soldiers. I have Cozens Hardy to help and two keen and capable Sergeants, Hicks and Barry. Began preparing the ground today. It is a beautiful little French village in a narrow valley, off the Somme Valley. The steep chalk slopes act as good butts for our shooting … I think our experiment here will prove a success. We are eliminating all but the salient points to be taught, and, with the help of the Amiens camouflage experts, introducing new dodges for teaching the use of ground … Poor Matthews, to help in the training, he had one day hidden, disguised as a German. He never forgot how the natives with pitch forks gave chase!

But while it was all very well training men to be snipers, challenges nonetheless existed in utilising their skills effectively. Officially and perhaps indicating what Army Command thought of the importance of sniping in general, the strength of a battalion sniping section was fixed at eight men, with an officer, a sergeant and a corporal in charge. The subaltern was originally 'Sniping Officer', but became 'Intelligence Officer' as it was increasingly vital that he interpreted the men's observations correctly and sent clear reports to Brigade Intelligence HQ.

In practice, the number of men selected for sniping depended on the attitude of the battalion commander. Albert Huxford said that:

> When we got back from the [sniping] course, we were told we were on fatigues and we worked like bloody navvies for the next week carrying wire and all sorts up to the lines at night. We got back and were put straight on stand-to in the mornings but we were also supposed to occupy our sniping posts during the day. It wasn't possible, and it came to a head when our Major found two snipers asleep in their post … they were put on charges but Lieutenant Gibbs, our sniping

> officer, stuck up for us and told the divisional Commander about all the work that we had to do. There was a real hue and cry about it and our Colonel got a rocket about misuse of trained men. We were left alone after that, but until the Colonel left the battalion, we were always out of favour.

There was also dissent in the way in which training should be undertaken. Herbert McBride, for example, recorded that he had read:

> Major Hesketh-Prichard's book 'Sniping in France' and find it exceedingly interesting. He describes at length the methods used in training snipers in the 'Imperial' schools. I also had the opportunity, during the summer of 1918 while on duty at Camp Perry, of observing the sniping school there, which was conducted by Major Godard – who had received his training at one of the above mentioned schools. It was all intensely interesting and exceedingly clever; the way they worked up artificial cover – camouflage and all that – but I could not then, and cannot now agree that their system was sound. For use in permanent locations – perhaps yes, but for troops that are continually shifting – no. It requires too much paraphernalia.

McBride also offered a personal opinion that:

> The true sniper must 'travel light' in order to be able to take advantage of all natural cover in whatever situation he may find himself. That, I think, defines the point I am trying to bring out – that to train men to the belief that they will always have a 'sniper's robe,' long grass and turnip tops, or other similar material with which to construct concealed positions – is to lessen their ability to take cover where none of these commodities are available. Furthermore, when out in the open, as in an advance, you do not have all day in which to get a position fixed up, or to get in position, or even to take a shot. Conditions change very rapidly and the sniper must have been trained to act and shoot at once – or pass up that opportunity.

The make-up of sniper sections was rarely recorded, but, according to Martin Pegler in *Sniping in the Great War*, the war diary of the 1st Civil Service Rifles (the 15th London Regiment) noted that, in June 1916, the sniping section was 30 strong, being 3 per cent of the strength (of the battalion) drawn evenly from the battalion. Three men including the sergeant instructor were pre-war regulars, the others having joined in or after 1914. Interestingly, the report goes on to reveal that the average age was 22 and that eight men were married. Half were civil servants, the others being clerks, merchants, commercial travellers, engineers, a tailor's cutter, a gardener, a waiter and a student. By 1918, however, five were dead and 11 had been wounded.

Frederick Crum suggested in *Memoirs of a Rifleman Scout* (1950) that the sniping establishment should comprise at least 16 men plus officer, sergeant and corporal. But he added a rider that 24 or even more might be required. A list of suitable spare men should be kept by the sniping sergeant and arrangements made to train them as occasion arose. These men could be used as company snipers when required.

From late 1915 onwards, ever-increasing acceptance that the sniper was part of the regular establishment of a front-line battalion was to have a profound effect on the German sniper's dominance of No Man's Land. In areas where British snipers were actively working, British casualties began to drop dramatically. An indication of the effect this had can be gleaned from a letter sent in February 1916 to Hesketh-Prichard by an officer serving with the 70th Infantry Brigade: 'We have only been in the line eight days and have got more Germans (certainties) than in the month before we came out.'

The Germans were so stunned by the British efforts that they all but ceased sniping in many parts of the Western Front. Infantryman Edmund Gottlieb wrote that:

> On the Somme, we had no problems with British sniping until perhaps May of 1916 when we were faced by a new regiment of Welshmen [probably the Royal Welsh Fusiliers] who had some very excellent shots. We began to suffer very greatly from their shooting and it was here I was wounded by a bullet that went through my cap as I was looking through my binoculars. It was good for me though, for I was in hospital through the Somme battles of July and August so I was also very lucky I think.

Private Thomas Durst, of the King's Royal Rifle Corps, had qualified as a marksman on the ranges, and on his arrival in France was 'volunteered' by his sergeant for sniper training:

> I was told off to report behind the lines to a new training school where I should be taught to become a sniper. I was unsure about exactly what that entailed, but there was no arguing with the sergeant. If he said 'go', you went. I had to hop lorries to get there and met up with a

> few other fellows from the regiment who had all been volunteered for sniper training. We wondered what it was all about but it was good to get away from the front lines for a while as well as the prospect of receiving an extra two bob a day specialist's pay.

Frederick Crum added in *Memoirs of a Rifleman Scout* that:

> As a science the art of sniping grew steadily, thanks to the efforts of a limited number of experts, who would not be denied. Gradually, from their constant watching, snipers grew into expert observers, and often the reports they sent in were of the greatest value. This led to their being specially trained and employed, and worked on a system, as observers. Later, their enterprise leading to successful work on patrol at night, they became instrumental in bringing about a scientific training in night work which eventually reduced the terrible wastage of men, sent out with every possible blunder of management, to reconnoitre and stalk the Hun in the dark. With regard to actual sniping, some of the tales in a book called *Sniper Jackson*, by Frederick Sleath give some idea.

Scotsman Frederick James Sleath had graduated in the summer of 1914 from Edinburgh University, where he had been a member of the Officer Training Corps. Sleath duly enlisted, being posted to the Royal Scots and then commissioned. After serving for some time in the Ypres sector as a sniping officer, a freak accident ended his front-line career. Published in 1919, his book was a largely autobiographical episodic novel. Crum then continued:

> Wherever the Hun was scientifically tackled we beat him every time, and more than once we left trenches, which we had taken over with an evil reputation for casualties, without a single mishap from hostile snipers. It was sometimes enough to kill a single really troublesome Hun sniper to secure complete moral superiority. In one sector, I remember, on our arrival, it was unsafe to show your little finger. When we came away, three weeks later, I saw one of our men coolly lathering his face in full view as he did his morning shave.

Many of the teachings of the leading sniper trainers of the First World War disappeared almost as soon as the Armistice had taken place. Excepting in the USSR, therefore, much of what had been lost had to be re-learned – often at great cost in men and equipment – when the Second World War began.

In Britain, even after the immediate post-Dunkirk crisis had passed, the threat of German invasion receded and the focus had switched to North Africa, there were still no sniper-training facilities. 'After Tunisia and Sicily, however,' Clifford Shore noted in *With British Snipers to the Reich*:

> The position was totally changed, and one of the dominant cries was 'Sniping'. Sniping became a War Office priority, and every assistance was given from the High Places.
>
> One of the objects was to produce instructors in every Battalion in the formation which was to become universally known during the European campaign as 21 Army Group, in order that a trained body of snipers was in each unit before D Day. And this was achieved mainly due to the efforts of … Majors Wills and Underhill, although the latter characteristically gives all the credit to Major Wills.
>
> The first school under the new scheme of things was at Llanberis in North Wales … from September 1943 to March 1944. This region is the wettest in the whole of Great Britain and the work was carried out in almost insurmountable weather difficulties.

The two-day course for battalion commanders was particularly beneficial:

> because if the Commanding Officer of a formation is disinterested there cannot exist in that unit any sniping in its true sense. But if the OC is keen on the game then sniping thrives; so do the snipers and there are many scalps on the sniping totem. It is essential … that a lead be obtained from the top.

Sniper-training facilities existed to serve British units in places such as Italy, the Lebanon and Palestine, but results were inconsistent. Clifford Shore wrote critically of efforts in Italy, where even though the sniping officer 'knew his job thoroughly … the sniping school never occupied the niche that was carved for it in Normandy and Holland; obstructionism at its worst was encountered'. He also damned the training attempted at British Army's Mediterranean Training Centre at Cassino, where:

> Sniping was a very small peg in a large hole. At all other centres of instruction everyone was very enthusiastic when put into the picture about sniping, but here things were different and every stumbling block was placed in the way of someone wanting to get things done. The men at the head of the school thought they knew a lot about

> sniping, and in reality knew very little, and as is usual with this type of individual they were resentful of suggestions.

The school was 300 miles or more behind the front line and, said Shore:

> was completely out of touch. And all the efforts to get the staff sergeants up to the line for a shot or two were useless. The only possible way for the officer concerned to get up to the forward areas was to go there on leave! Every unit in the line was crying out for assistance and advice, and some of the sniping instructors were sobbing out their hearts to respond to the plaintive cries and yet a relatively minor officer type, of a species occasionally found in all Armies which goes out of its way to obstruct rather than help, stood obdurately in the way. It is ironical to think that these obstructionists were often decorated for their services!
>
> The average sniper in Battalion was a lost sheep, and one could have helped him a great deal, and when that help was possible on those too-brief trips not only were the snipers themselves greatly appreciative but so were their Company commanders and COs.

Conversely, the personnel of what was to become 21 Army Group Sniping School landed in Normandy on 17 August 1944 and soon began building a range near Courselles sur Mer. An instructors' course was in place by the end of the month. Eventually moving into the Nijmegen salient, the school was recreated only a few miles behind the front line and finally given coveted Army Group status on 24 November 1944.

Affairs in Germany generally followed the progression in Britain: from localised improvisation to standardised training. Bruno Sutkus described in *Sniper Ace* how, during his first session with live ammunition:

> At one hundred metres on the twelve-ring target I scored four twelves and an eleven. Next I had to fire five rounds at a camouflaged target and scored three twelves and two elevens.
>
> I did not stay long in the training company. After a month or so I was drafted to the Sniper School at Vilnius in Lithuania. The school was located in a barracks near the St Peter and Paul Church. The same complex housed an officers' training establishment. Our course lasted from 1 August to the end of December 1943.
>
> We were shown captured Russian film from which we gained an idea of the things a sniper had to master, especially fieldcraft, concealment, marksmanship (of course) and range estimation. Over

> these five months we learnt to absorb all the minute details a sniper had to bear in mind in order to spot the enemy hidden in the natural environment and to avoid being picked out himself. The instructors were good. In the countryside they taught us how to recognise a target, pass information, estimate range and shoot at moving targets. I developed great accuracy in the latter.
>
> I realised that in those five months I had to absorb what I could in order to survive in the field. At the end of the course all those who qualified received the most modern sniper's rifle with telescopic sight, binoculars and a camouflage jacket (*tarnjacke*). I was given a certificate confirming that I had passed out from sniper school and was warned never to allow another person to handle my rifle.

The Americans had to re-learn not only the value of sniping but also ways in which techniques could be taught. Among the principals were US Army's Claude Harris and George Van Orden of the Marine Corps, both successful competitive marksmen. Sasser and Roberts remark in *One Shot – One Kill* that:

> The sniper school that became the model for future schools in both the Army and Marines sprang up from the hills north of San Diego, California, in January 1943, The first commander of the Marine Corps Scout and Sniper School at Green's Farm was USMC Lieutenant Claude N. Harris, who seated his sniper recruits underneath the eucalyptus trees in the front yard of the old farmhouse turned school headquarters and growled, 'Snipers can save a country. Look what they're doing for Russia'.
>
> The Corps brought him back from a combat outpost in the Pacific to organize and command the new academy for killers. Rugged but mild-mannered, appreciating the value of the single well-aimed shot, he became the ideal boss for the gangs of expert riflemen his school turned out every five weeks.
>
> Saying it wasn't enough for a sniper to be able to plink a round into a stationary target at four or five hundred yards, Harris built upon the experiences of previous snipers like Captain McBride to mold a human weapon who could shoot well, but who could also find his way around on the battlefield and survive.

Sasser and Roberts also draw attention to the recollections of Army Sergeant John Fulcher, a part-Cherokee farm boy from Texas, who described the often-impromptu nature of sniper training post-Pearl Harbor:

> You had to volunteer for it though. Somebody in ranks muttered that the life expectancy of a sniper in combat was about what you might expect of a snowball dropping in west Texas in July. Whoever said it just stood there in his khakis and stared straight ahead. He wasn't volunteering.
>
> A couple of boys stepped out of ranks. What the hell. I could do it. I gave a shrug and moved forward. The sergeant grinned at me. 'I figured you would, Fulcher,' he said.

Fulcher was fortunate that the sniper school was held at the same camp, reflecting that at the times when trainees:

> weren't running, making forced marches, or stalking each other out in the mesquite, we were shooting on the rifle ranges. A lot of shooting at targets about five hundred yards away. They issued us new M-1Ds still in cosmoline. The D model was an ordinary Garand, except it had a heavy match barrel and flash suppressor, a set action for the trigger so you could get a smoother squeeze and discharge, a leather cheek piece, and a telescopic sight.
>
> The scout/snipers were supposed to operate in six- and twelve-man squads, pushing out ahead of friendly elements, into no-man's-land and sometimes even behind enemy lines. We provided security by being the eyes and ears for the battalion. Whatever the enemy was doing, we found out about it and funneled back the information.
>
> That was the scouting part. The other part was the skirmishing. Wherever we found the enemy – his patrols and squads and lookouts – we picked him off until he was afraid to go where we were.

However, Claude Harris's course was much more comprehensive, seeking to give instruction in using a compass, reading maps, interpreting aerial photographs and sketching what could be seen. Effective use of camouflage, making satisfactory hides and how to master concealment techniques were taught alongside stalking and long-range shooting over a five-week period. As Sasser and Roberts note in *One Shot – One Kill*:

> Marine students scattered into the surrounding canyons where they crawled through barbed wire entanglements, dug hasty field fortifications, made range estimates, and drew maps from memory. They did a lot of scouting – 'silent walking' – over rough terrain. One of their tests was to lie in total darkness and estimate the number of men in a passing patrol by the sound of their footfalls.

> In the mornings and afternoons, using 1903 Springfields and modified Garands, the students fired at man-silhouette targets set at various ranges up to five hundred yards. Snap-firing at disappearing and moving targets managed by wires, they learned how to hold at any range. The final week was devoted almost exclusively to telescopic long-range firing.

Keen observation and good range-estimation have always been vital to efficient sniping, and even the earliest training programmes laid emphasis on the problem. Good observation paid vital dividends. As Vasiliy Zaytsev reflected in *Notes of a Russian Sniper*:

> We lay in silence, observing. The rays of the setting sun were lighting up the hill. They illuminated all the dark recesses and bathed every outcrop in bright relief. A bit further down the hill some spent German artillery shell cases lay scattered. I had nothing to do, so I counted them. There were twenty-three. Then I did a double-take. One was missing a bottom!
>
> Through a shell case like that, just like through a telescope, someone could see a long way into the distance. I raised myself a little. Suddenly there in the casing – it was like a flint struck a spark! An explosive bullet ripped into the embankment behind me.

And Evgeniy Nikolaev noted in *Red Army Sniper* that, by 1942, the practice of sniping was becoming more and more widespread in front-line units:

> Among its initiators were snipers Vladimir Pchelintsev, Feodosiy Smolyachkov, Ivan Vezhlivtsev, Pyotr Golichenkov and others, all of them well known across the entire front. The front-line newspaper published their letters and described their careers and activities in the forces. 'Here they are, our glorious patriots,' wrote the newspaper. 'Their sniper tallies are already over a hundred. Corporal Ivan Vezhlivtsev has already eliminated a hundred and thirty four Nazis, Corporal Golichenkov – a hundred and forty, Sergeant Pchelintsev – a hundred and two. Praise and honour to the avengers of the people. Emulate them, Red Army soldiers!'

Vladimir Pchelintsev is best known for accompanying Lyudmila Pavlichenko on a tour of North America and Britain, undertaken in 1942 as a propaganda exercise seeking to increase support for the Russians. But he was also a very successful sniper. A letter from Ivan Vezhlivtsev to Pchelintsev, quoted in the Nikolaevs' book, stated that being a sniper:

> takes a lot of tenacity, stamina and composure. My own first day out stalking was not altogether successful. A Nazi appeared on the cutting I was observing. I fired and missed. But why? It was my own fault – I was in too much of a hurry; I was afraid the German would get away. Several hours later at the same cutting I felled first one Nazi and then another.
>
> Since then it has been my rule, as soon as the target appears, to aim precisely, quell my nerves and make the shot calmly. Patience and calm are the main things in our business; firing facility depends on how correctly the distance to the target has been determined and how precisely the sights have been set.
>
> I have a number of ready equipped firing points in various places. This is essential, because the enemy also carries out observations, seeks out our snipers' nests and, upon discovering them, opens fire with mortars or even cannons. When the Nazis open fire, I change my site.
>
> You need to remember that the enemy is crafty and resorts to cunning. A few days ago the Germans put up a dummy to draw my fire. It was wearing a helmet and a white coat. Through my telescopic sights I immediately detected the enemy ruse and did not waste any cartridges.

Vezhlivtsev then observed that 'love for one's rifle is everything':

> You must protect your rifle more keenly than your own eyes. Otherwise it might let you down. I clean my rifle after every firing session, meticulously care for its optical devices, put the cover on them, and rub the glass with a special brush. Such are my tips for snipers. I have my experience and others have theirs. Let's swap it, make use of it, and broaden the competition between us. There must be no mercy for the enemy!

Another great virtue was the ability to co-operate. Many of the snipers employed during the early period of the First World War worked alone, but, as conflict broadened, the value of 'spotters' to share the burden of continuous observation – and to keep the eccentricities of individual marksmen within bounds – came to be appreciated.

But by the time of the Second World War, snipers habitually operated in pairs. This method allowed one to shoot while the partner observed, occasionally exchanging roles to reduce the effects of boredom, eye-strain or fatigue. Evgeniy Nikolaev, for example:

> operated side by side with Ivan Dobrik … until he was seriously wounded in July 1942. He suffered a serious head injury from an enemy shell splinter in the rear of our regiment. It was not at a commander's behest that we went out hunting together, but rather that we sought each other's company. At the front Ivan and I were drawn still closer to each other by our joint operations, which had a single goal: to wipe out the Nazis – whenever, wherever and however we could, just so long as more of them were annihilated …
>
> It was a pleasure to have him for a friend; he was a real comrade – straight-forward, responsive, honest and kind. The only thing was that he could be unreasonably hot-headed at times; on occasion he could dish it out to his superiors in a way that made your ears burn! But he even got away with that; presumably, he must have had some good handlers …
>
> As marksmen, Ivan and I were on the same level, but there was not a trace of competitiveness between us. On the contrary, each of us took pleasure in the success of the other, helped him, and shared his own knowledge and experience when necessary. Different in nature, Dobrik and I somehow complemented each other …
>
> We learned to understand each other from first glance, from the first gesture. I knew how to calm down my unreasonably temperamental and thus not always prudent friend. And if my partner suddenly thought up a dangerous 'escapade' in a duel with a German sniper, my arguments and suggestions never drew a hostile reception from Dobrik; weighing everything up and analysing it, he would accept a different, proper, decision. In duels with the enemy he always emerged the victor. No bullet from Ivan's rifle was ever wasted.

Klavdiya Panteleyeva affirmed that when she entered sniper school:

> Marusiya Chikhvintseva and I stood next to each other, so we remained partners. Many of us graduated, I couldn't say how many now, but they sent us to all fronts. My group had about 12, six pairs. Simultaneously. A squad was 10 soldiers, but there were more of us. Partners were together all the time, often at arm's length.

However, time often passed without incident. Andrey Ulanov, in his forthcoming *Stalkers of the Enemy*, notes that Anatoly Navara recalled:

> We drew lots to determine who would take the first shot. When it was my turn, I would adjust the sights, looking for a target, while my partner would just indicate significant landmarks where he might have noticed something. In order to be able to record in my sniper booklet that a target had been hit, I needed the confirmation of my partner and a lookout from the infantry; so, not everything got recorded … We took very few shots in the course of a day – one or two, three at the outside. There were days when we didn't fire at all.

TAKE YOUR WEAPONS!

> I am indebted to my weapon for my initial success. A rifle is a soldier's best friend. Treat it with care and attention, and it will never let you down. Protecting your rifle, keeping it clean, eliminating the slightest faults, greasing it moderately, regulating all its parts and adjusting it properly – that is the attitude a sniper should have towards his weapon.
>
> At the same time it does no harm to be aware that, for all their standardized parameters, there are in principle no identical rifles. As the saying goes, each one has its own character. This character may manifest itself in the tautness of the various springs, the ease with which the bolt slides, the weight of the touch required on the trigger, the state of the bore, its degree of wear and tear, etc. On a number of occasions I have returned from the 'hunt', hungry and shivering with cold, and the first thing I have done is to set about cleaning the weapon, putting it in order. For a sniper this is law.

So wrote the renowned Red Army sniper Vladimir Pchelintsev in *Kak ya stal snayperom* ('How I Became a Sniper'), published in Leningrad in 1942 and detailed in the Nikolaevs' book. Sniping memoirs contain thousands of words about their individual experiences, but few other than Lyudmila Pavlichenko's *Lady Death* pay detailed attention to the suitability of rifles and sights. However, Bruno Sutkus was among those to draw attention to their importance:

> The rifle had to be cared for; the rifling needed to be perfectly clean and kept well oiled. One's life depended on this. The weapon had to be hung up and so stored so that it would not fall or be damaged. One always had to be able to rely on one's rifle.

US Army Sergeant Bill Jones was more than happy with his allocation:

> I got volunteered to be the company sniper. We only had one for each company. My platoon sergeant sent me down to the armorer who signed me out a factory-new, bolt-action .30-caliber 1903 Springfield with the finest scope – one of the *only* scopes – this ol' son of an Appalachian sharecropper ever saw. The armorer grinned.
>
> 'It ain't a woman, Jones, Don't stroke it.'
>
> 'I ain't never seen a rifle as purty as this one.'
>
> 'Well, you take that rifle, Jones. It'll kill some Germans for you when you get over there if you'll take care of it.'
>
> Everyone else in the company had either the standard issue M-1 Garand or the shorter .30-caliber carbine. Because of the *rifle*, everybody knew who I was. I was *the sniper*. That was how I thought of that relationship too – *the rifle*; *the sniper*.
>
> Because we were hill people when I was growing up and had to depend on eating squirrels and rabbits and possums to survive, I had become a fair shot with a .22. I got even better with *the rifle*. After I got it zeroed in just right, I could hit anything I wanted. I knew what elevation I needed for a certain range and I could find it by the clicks on the scope, day or night. I might have been slower than the other boys with their Garands, but I knew I was a great deal more accurate. I could hit what I shot at.

However, not everyone was as pleased. In *A Rifleman Went to War*, Herbert McBride somewhat cynically observed that:

> We learned that a rifle can be very much like a woman and that the lovable sweetheart of pre-nuptial days may fall far short of being the trustworthy and reliable helpmate in wedded life.

Optical Sights

Though Whitworth cap-lock rifles fitted with Davidson telescope sights had shown their effectiveness at long range during the US Civil War, and though improved sights had been tested at the end of the nineteenth century by many armies – particularly in the USA – no purpose-designed sniper rifle was being issued anywhere when the First World War began.

In 1914, however, very little attention was being paid to sniping, as lessons learned from the American Civil War, the Second South Africa (Boer) War and the Russo-Japanese War were all but lost. The slate had been wiped clean.

Consequently, the havoc wrought on the Western Front by German marksmen came as a shock not only to the British and French field commanders but also to the rank-and-file.

Rapid stagnation of what many people had expected to be fluidity of battle lines, with the creation of hundreds of miles of trenchscape, promoted a style of defensive shooting for which few men were prepared.

The optical sight proved to be at once the sniper's greatest asset and biggest liability. New Zealander Private L. James recalled how he was:

> sent on a course after I returned [from Gallipoli] and we learned how to use telescopic sights. I had never used one before, always iron sights for my sniping and I must say I never took to them. They misted up and never seemed to keep their zero, but they did prove useful in the trenches as you could see Jerry's loopholes much better. I still preferred to shoot with open sights, as I found snap-shooting was much quicker with them.

Problems of supply restricted the distribution of optical sights for some time. Frederick Crum noted in *Memoirs of a Rifleman Scout* that:

> It was not till later that it began to rain telescopic rifles, so fast that Major Sclater's and Hesketh-Prichard's Schools could not keep pace with instruction in how to use them. In those days we enthusiasts spent large fortunes privately on telescopes, periscopes, elephant guns, etc. We were the pioneers who proved the need of some sound system and of official sympathy; and, as is always the case with pioneers, it was an uphill and expensive task.

Herbert McBride highlighted the dichotomy in *A Rifleman Went to War*, observing that as he had never advanced 'across a battlefield and act the part of a modern rifleman while armed with that Ross rifle and Warner & Swasey telescope' he could not give:

> any actual experiences as to just how the telescope sight would work under such conditions. I do not think it would stand up under the necessary rough handling but would soon get out of alignment for real accurate shooting, the only sort of shooting that is worth anything, by the way.
>
> But, supplemented by the regular iron sights, it would be all right and even if the telescope did get out of alignment or have its

> cross hairs jarred loose, it would still be splendid to have along for purposes of observation. One could easily look through the misaligned telescope for a target and then hit that target with the iron sights once it was located.

Magnification, field of view and light-gathering capabilities dependent on the lens system had been debated. A degree of consensus had been reached by 1914, but adjustment of optical sights was problematical, and mounts were often not rigid enough to ensure that the telescope stayed in place from shot to shot. Indeed, some of the target sights pressed into service were spring-loaded: the Winchesters of the First World War, for example, often had to be pulled back into place before each shot. Herbert McBride wrote that:

> There must be some arrangement for changing both elevation and windage, moved by means of sufficiently large devices to take hold of, showing settings in both plain figures and 'clicks.' I want no range-finding contrivances, etched lines, battle slogans or anything of that sort on the lens either – what is mostly needed is the largest possible entirely clear and distinct field of vision with a definite aiming device in it – and nothing else.

But there was much more to the story than mere technical detail. The widely perceived success of the Turkish snipers operating at Gallipoli, allied with a widely held view that the rear-locking Lee–Enfield was not as accurate as the front-locking Mauser, initially led to some Turkish rifles being turned by Anzac and British marksmen against their erstwhile owners.

Attempts were soon being made to procure optical sights, but these were in short supply even on the Western Front in the spring of 1915. Consequently, the Lee–Enfields used for sniping at Gallipoli were usually open sighted. Some men used their own competition-type aperture sights, but long-term solutions were clearly needed.

Among the most popular expedients were the 'Galilean sights', two widely spaced magnifying lenses in a holder of some type. The 'Sight, Optical (Martin), Mark I' was approved on 1 May 1915, though apparently preceded in issue by the 'Sight, Optical (Lattey), Mark I' and 'Sight, Optical (Neill), Mark I' – also known as 'Barnett' or 'Ulster' – which were not formally approved until 28 September 1915. War Office contract books reveal purchases of 14,125 Lattey, Neill, Martin and Gibbs patterns prior to 7 July 1915.

Private purchases may suggest that 50,000 Galilean sights saw service. A Lattey in the collection of the Australian War Memorial is known to have been used at Gallipoli by Fremantle-born Trooper Charles Livingstone of the 6th Australian Light Horse, and there were doubtless many others. However, the widely separated lenses allowed dust, reflected light and radiant barrel-heat to degrade performance.

Magnification was generally no more than 2.5×, and the field of view was usually only about six feet at 100 yards. So, despite the manufacturers' claims, few marksmen expressed favourable opinions: Galilean sights, easily damaged and just as easily lost, were no match for the conventional telescope. Consequently, the failure of alternative ideas allowed optical sights to regain pre-eminence even though some men remained sceptical – particularly if rapid shooting was needed at comparatively close range. Herbert McBride opined in *A Rifleman Went to War*:

> I liked to have the iron sights in position on my rifle, and fixed so either they or the telescope could be used. Then, when conditions prevented the use of the scope, or it got 'out of whack', I went right along from where I was at the time, and used those iron sights. Have issued a second rifle for iron sights only, do you say? Listen here, you, I left that rifle back in the dugout this morning, and we're out here in this shell hole for the whole damn day and can't get back until after dark.

British War Office records show that 2,914 telescope sights were purchased from 'The Trade' from 1 August 1914 to 4 July 1915, but they came in such variety – some were German-made – that repairs proved to be impossible in the field. For, as McBride said:

> The telescopic sight needs a lot of attention, care and babying, to keep it in proper condition and to make it 'stay put' … One of the main reasons why I liked so well to shoot from Sniper's Barn was that while in there you could take care of and handle the telescope right. For one thing, when through firing I would leave the rifle (and the big observation telescope) right there, hidden away under some old bales of tobacco, and there was no danger of its being dropped and the sight jarred out of alignment. I had several wide strips of old woollen blanket and used these to wrap around the entire telescope and action when I put the rifle away. In there I could keep the lens dry and clean.
>
> You cannot work a telescope in the rain with any degree of satisfaction, and even a light fog or mist will cloud up the lens continually. The rifle cannot be carried through wet or dew-covered

underbrush without throwing water all over the lens. Even the taking of the rifle from a comparatively 'warm' dugout out into the open air of a cool or damp day may fog up the lens so badly you cannot see through them [sic]. Often on a misty morning one is constantly wiping off the lens in order to see at all.

And finally there is the dirt and the mud to be considered, and the wiping of a few specks of mud or grit from the lens of a telescope is a job to be gone about with much caution – and a clean soft rag, else you will soon ruin the glass from scratches.

Rifles for Snipers, 1914–18

The combatants carried bolt-action rifles, generally with late nineteenth-century origins. The British had the Lee–Enfield, which was the front-line weapon of all British colonial forces, except Canada – where, attempting to reduce reliance on supplies from Britain and play to national identity, the straight-pull Ross had been issued.

Experience in the Boer War showed the Lee–Enfield to be cumbersome; after lengthy trials, therefore, a potentially revolutionary short-barrelled rifle intended for universal issue was adopted in 1902 – even though the pattern was not formally approved until January 1907. The 'Rifle, Short, Magazine Lee–Enfield' (SMLE), just 44½in overall, weighed 8.15lb empty and had a detachable box magazine holding ten rounds.

The French had the Lebel and the Berthier, the Austro-Hungarians had Mannlichers, the Italians had the Mannlicher–Carcano, while the Belgians, Germans and the Turks had Mausers.

Excepting the straight-pull Austro-Hungarian Mannlichers, the rifles all relied on bolts which turned so that locking lugs were disengaged: strong and comparatively simple, though considerable differences existed in the construction and performance of each type.

When war began in 1914, the SMLE was the only compact universal-issue rifle being issued in Europe: the French, the Germans, the Italians and the Russians were all using cumbersome full-length infantry patterns and a variety of short-barrel carbines and musketoons for special purposes. Herbert McBride concluded in *A Rifleman Went to War* that 'for effective use as a rifle in battle' it had to be:

> just as compact and 'handy' as it is possible to make it and still retain accuracy and other punch. It is probably not possible to build a high

> power, bolt-action rifle that would be as handy as the little .30/30 carbine or 'saddle gun', but that is my idea of what a handy rifle should be. No, with the bolt action and box magazine, it would be impossible to get the 'balance' just the same and that has a lot to do with the 'handiness' of any rifle. Perhaps the Springfield Sporter could be worked down somewhat. It is not bad, just as it is, yet even it would have appeared 'clumsy' beside that little Mauser I took that day.

The unprecedented rate at which fighting accelerated highlighted the lack of equipment. As far as sniping was concerned, the first solutions, particularly in Britain, relied heavily on the gift or impressment of sporting rifles while development of specialist equipment began and large-scale manufacture of optical sights was pursued: a programme which would clearly take time to resolve, owing to the inability of the British optical industry to supply sufficient sights at the outset. Eric Penberthy wrote in *The English Review* in 1920 that:

> Many privately-owned weapons began to appear in the trenches – sporting rifles of various patterns and calibre. By the irony of fate, one privately-owned rifle, fitted with a telescopic sight, which did great execution in the hands of a sniping officer of the Staffords, was a pre-war gift by the ex-Kaiser to a well-known British General.

Adding that, 'on Hill 60, in 1915', he remembered:

> A young officer who was not altogether innocent of 'eye-wash' telling me that he had killed several Huns by the simple method of firing through the enemy's sandbag breastworks with a Rigby express rifle. He could not see his targets, so I asked him how he could tell when he scored a hit. 'Oh', he said, 'those trenches are very wet and I hear the bodies splash when they fall into the water'.

And Herbert McBride related in *A Rifleman Went to War* how, though he had never seen optical sights during his training, while 'scouting … in back of our lines opposite Messines Ridge' in September 1915 he had run into a sniping post manned by an officer and two snipers from the Buffs (East Kent Regiment):

> I naturally stuck around. I stayed there for a couple of hours, but what I saw being done did not get me any too highly excited over that type of sniping.

> These fellows from the Buffs were using the ordinary, short Lee-Enfield rifles, upon which they had mounted telescopes made by 'Stanley-London.' These scopes were short, brass tubes, about ten inches long and three-quarters inch diameter; they had a device for changing elevation, but no method of making lateral corrections, and for windage you simply had to hold off. I do not happen to remember the power, but recollect that the field was very limited although visibility was excellent ...
>
> This outfit they were using may have been very good for reasonable ranges, but these chaps were so far in back of the line that it was hopeless for them to think of doing any definite or accurate shooting. The nearest enemy targets were at least eleven or twelve hundred yards away, but they were doing most of their shooting at targets well over on Messines Ridge and I knew the range to these points was about two thousand yards. Now, with a machine gun, it is possible to put a burst on a target or group of men at such ranges and break them up and possibly hit one or two. But my personal experience has been that the firing of single shots, at individual targets, at ranges of more than 1000 yards is just a waste of time and ammunition. I have tried it myself and seen many others try it, but never saw any indications that the target had been hit.

Another short-lived idea was inspired by the trench periscope. In March 1904, William Youlten had been granted a British Patent protecting a periscope sight attached to the left side of the SMLE fore-end with an offset head to place an angled mirror behind the centreline of the back sight.

This device led to the 'Rifle Butt Hyposcope', which was made in small numbers during the early part of the First World War. A prismatic sight hood, attached to a pivoting lever on the side of the butt, could be aligned with the sights while firing from cover.

Self-contained periscope rifles were used on the Western Front, principally by the French, and in Gallipoli by Anzac marksmen. They had the advantage of protecting the firer from counter-sniping, but were generally somewhat difficult to control.

Though there is at least one other Australian claimant and patents for similar devices had been filed prior to the war, General Sir William Birdwood recorded that 'complete moral superiority over the Turk is partly due to a very clever invention of a man named Beech, who produced a periscope rifle.'

Lance-Corporal William Beech of the New South Wales 2nd Battalion had simply fitted an SMLE in a skeletal wooden frame, relying on mirrors to provide a periscope sight. Charles Bean observed in the *Official History of Australia in the War of 1914–1918* that:

> On May 19th Major Blamey, going round the front trench of the Pimple during the last hours of the Turkish attack, had observed two men of the 2nd Battalion [with a wooden frame] attached to a rifle, which they were endeavouring to lay on the parapet. 'An arrangement so that you can hit without being hit,' one of them explained. It was a device for aiming a rifle by means of a periscope so fixed that the upper glass looked along the sights, while the sniper gazed into the lower one. Blamey believed that the device might be valuable, and the inventor, Lance-Corporal Beech, was afterwards brought to headquarters to apply it to a number of rifles.
>
> By May 26th a factory was started on the Beach. The first periscope rifle had been taken into Quinn's on the previous day … It was tested at many posts, and at ranges up to 200 or 300 yards was found to be an accurate and deadly weapon. From that time it was constantly used by the Anzac troops wherever the trenches were close. At Quinn's, where previously the garrison could scarcely fire a shot by day, it helped to beat down the enemy's sniping, so that it became possible to fire from the loop-holes and even, for a few seconds at a time, to look over the parapet.

Bean also suggested that large numbers of Beech-type periscope rifles were made, naval vessels being stripped of glass which could be used to make the periscope sights while frames were made from packing boxes. Guns were used in small numbers by British units in southern Gallipoli, and by the Turks who captured and then copied them. Ion Idriess related how:

> Parts of the trench were so close to the enemy that they could not use their loopholes in daylight for fear of attracting a fusillade of bullets for their trouble. The periscope rifle was ideal in these sections of the front as it was very accurate and easy to use over short distances up to around 150 yards.
>
> For longer range sniping, the periscope was often used by the observer to locate targets with the firer having to 'snap shoot', adopt a firing position, locate the target and fire, all within about three seconds.

Finally, on 4 May 1915, an optically sighted version of the Mk III SMLE was approved, individual guns being selected for transformation only if they were capable of two-inch groups at a hundred yards.

Contracts were negotiated immediately and, by the close of hostilities, 8,933 had been delivered. Most of them had optical sights made by the Periscopic Prism Company or Aldis Brothers, though 907 had Winchester A5 sights. Excepting some Winchesters placed on the centreline, the British sights were considerably offset to the left side – attracting the ire of Hesketh-Prichard in *Sniping in France*:

> One of the greatest difficulties we had ... was the fact that the telescopic sights were set, not on top, but at the left-hand side of the rifle. This caused all kinds of errors. The set-off, of course, affected the shooting of the rifle, and had to be allowed for, and the clumsy position of the sight was very apt to cause men to cant their rifles. Worse than all, perhaps, in trench warfare was the fact that with the Government pattern of telescopic sight ... it was impossible to see through the loopholes of the steel plates which were issued, as these loopholes were naturally narrow ...
>
> Why the telescopic sights were set on the sides of the rifles was never ... satisfactorily explained, but it was always said that it was done so that rapid fire should be possible ... then surely whoever was responsible can have had no knowledge whatever of the use of telescopic sights.

The SMLE was then replaced by the Pattern 1914 Mk I* W (T), approved on 11 April 1918 with a P/18 sight mounted centrally.

Once the First World War began, realising that Lee–Enfield production could not meet ever-growing needs, the British placed contracts for P/14 rifles with the Winchester Arms Company, the Remington Arms–Union Metallic Cartridge Company and the Remington Arms of Delaware Company. Three differing sub-variants were introduced to British service on 21 June, designation-suffixes being needed to distinguish them as the products of each factory differed sufficiently to prevent interchangeability.

The improved P/14 Mk I*, approved in December 1916, had comparatively minor changes in the bolt mechanism; and rifles with fine-adjustment back sights had designations with an additional 'F' suffix. Though the British authorities decided to curtail orders in the autumn of 1916, nearly two million P/14 rifles had been made when assembly ceased.

Adopted on 11 April 1918, the P/14 Mk I* (T) became the British Army's sniper rifle. Mounts for the P/18 Hensoldt-copy telescope sight (carried centrally above the bolt) were added on the left side of the breech and back sight protector, long-range sights being removed.

About two thousand rifle/sight combinations had been assembled prior to the Armistice, though few of what was perhaps the best sniper rifle of the period reached service before fighting ceased. Most of the survivors were discarded in the 1920s, though substantial numbers actually went into store. When British nomenclature changed in 1926, the P/14 Mk I* (T) became the Rifle No. 3 Mk I* (T).

Though optically sighted SMLE and P/14 rifles served British Army snipers with distinction, the Canadians initially preferred the controversial Ross: now widely regarded as a failure, it was withdrawn from the trenches in favour of the SMLE when it proved to jam with embarrassing frequency.

However, the Ross was extremely accurate and, as a result, widely appreciated by snipers. Ultimately, some guns, possibly owing to damage, were altered to half-stock form. This generally improved accuracy by providing a barrel that much more closely approached the target-shooter's free-floating ideal than a military design encumbered with bands, handguards and nose-caps suited to bayonet fighting. Herbert McBride, 'who had trained intensively … on the Barriefield ranges in Canada and at Hythe, England', noted that in the 'newly-organized Sniping School, near the village of La Clytte':

> I was issued a Ross rifle – one of the lot made for and used by the members of the Canadian Palma Team at Camp Perry, in 1913. These Ross rifles were exceptionally accurate and dependable with the Mark VII ammunition we were then using …
>
> With the rifle I got a telescope sight. It was one of the type – new at that time – made by the Warner & Swasey Company. Prismatic and mounted on the left side of the rifle, it might not rate so high now but, at that time, it was better than any other I had ever used. One of the best features was that it could be mounted and used without interfering with the iron sights. I had a little trouble in getting it securely mounted so that it would not jar loose but finally, by using a wedge – made of a piece of safety-razor blade – and salt water, got her on so tight that I came near being court martialled when I finally turned it in.

McBride then noted that 'each rifle had been fitted with a particular sight, and had been extensively tested':

> I happened to draw rifle No. 140 and sight No. 49. A very few shots fired at the improvised sighting range made me familiar with the scope adjustments and permitted me to check in the scope against the iron

sight settings. So, late in November 1915, I came back from La Clytte with as good and reliable a sniping outfit as was available at that time.

He considered the Ross to be thoroughly reliable and accurate:

> Even in the strenuous rapid-fire tests, where fifteen shots per minute were required, it never failed. During a competition, one man fired thirty-three aimed shots (all on the target) in one minute, while many others exceeded twenty-five shots. In accuracy, up to six hundred yards, at least, it equalled or excelled any rifle I had or have since fired – the Springfield not excepted.
>
> I have seen as many as forty consecutive bulls-eyes made with it at three hundred yards – on the six inch bull – and correspondingly good scores at five and six hundred yards. We had no regular practice beyond the latter range but, on several occasions, I have used this rifle successfully at ranges up to one thousand yards or more.
>
> Mindful of all this, it can well be understood that we went into action with all the confidence in the world in our rifles. Every man in the original battalion had fired hundreds, yes thousands of rounds, each with his own pet rifle, and knew just what he could do with it.

But McBride also had doubts:

> The first time we were called upon to repel a determined attack, and sustained rapid fire was in order, it was found that the Ross would not stand up under that kind of treatment. Wonderfully accurate weapon as it was, it was never built for fast, rough work.
>
> The first complaint against the Ross rifle was based on the ground that it was too long and unwieldy for satisfactory use in narrow trenches or when crawling over open ground where the cover was sparse. It was difficult to handle it so as to avoid exposure to the enemy – especially with the bayonet fixed. However, as the French Lebel was equally long, and their bayonet much longer, this point did not carry much weight with the higher command. But when they commenced to freeze up on us, it was acknowledged that the matter was serious. They tried all sorts of stunts to remedy the trouble, sending the rifles out back of the lines to the armourer sergeants, who reamed the chambers out larger so the cartridge would not fit so tightly, and all that, but it was no go, and the ultimate solution was to take the Ross rifle out of the trenches and issue every infantryman the regulation Short Lee–Enfield rifle.

When the American Expeditionary Force (AEF) arrived in Europe in 1917, sniping was still largely ignored in the US Army although limited quantities of M1903 Springfields with M1913 Warner & Swasey or Winchester A5 sights were available. Herbert McBride remarked in *A Rifleman Went to War* that:

> Now, I hate to say anything in disparagement of our Springfield Rifle (Model 1903 – with 1906 ammunition). It does not require my recommendation. Too many people know all about it. But, why don't they put a sight on it? As it now is, I would certainly pick one of the S.L.E. (Short Lee–Enfields) for the ordinary, short-range work of actual battle. Argue all you want to about ballistics but what a man needs when he gets into a fight is a short, 'handy' weapon – something with which he can take a hasty, snap shot at a target which only shows for the fraction of a second and then disappears. And he wants a sight that you don't have to hunt around for – just something that you look through – not look for.

A modification of the .303 P/14 chambering the ·30–06 round, the 'Enfield', had been adopted for service as the M1917 owing to a shortage of M1903 Springfields, but no specific M1917-type sniper rifle was made – even though the Enfield provided the basis for the 'Telescopic-Sight Rifle, Model of 1918'. Unfortunately, the M1918 sight, copied from a pre-war Goerz Certar, failed trials so comprehensively that the entire project came to a premature end.

There are comparatively few records lauding the performance of AEF snipers, though their numbers included skilled marksmen such as Alvin York and Herman Davis; Herbert McBride, though an American citizen, achieved his kills while serving with the Canadian Army.

York's exploits were once well known, thanks to a 1941 film starring Gary Cooper, but revolved around the largely single-handed elimination of a machine-gun post – using his open-sighted M1917 Enfield and a Colt–Browning M1911 pistol – and the consequent capture of 132 German soldiers. Though his story was questioned by others in his unit, battlefield archaeology has now largely confirmed its veracity.

Davis, who soon sank back into anonymity, was perhaps more of a sniper, achieving about 60 kills with an open-sighted M1903 rifle, including five German machine-gunners at a range estimated by observers to be as great as 1,000 yards. This Davis apparently regarded as 'just normal shooting'.

In Germany, despite a better manufacturing base than Britain or France, the military authorities were initially unable to equip their snipers effectively. Almost as soon as war had begun, therefore, *Jagdgewehre*, hunting rifles fitted with commercial Gérard, Goerz and Zeiss telescope sights, were pressed into service in small numbers.

The first issues seem to have been made in Bavaria as 1914 drew to a close. Viktor Amadeus, Herzog von Ratibor, then sponsored an attempt to collect as many sporting guns as possible, though there is no evidence that the 'twenty thousand' said to have been collected – this may simply have been the programme's goal – were distributed widely.

They ranged from bolt-action Haenels, Mannlichers and Mausers to break-open shotguns, rifles and even *Drillinge* and *Vierlinge* (three- and four-barrelled guns). But some recipients were more than pleased. For example, Herbert McBride acquired a sporting rifle which had been taken from:

> a young and cocky Yager who had been wounded and taken prisoner. That one was a beauty. Short and trim – a regular 'sporter' in fact. The former owner vouchsafed a supercilious smile when I held it up beside my own heavy Ross, and I don't blame him. He had a real, honest-to-goodness battle rifle, beside which ours were just clumsy clubs. I never had a chance to really give it a good try-out but did take a few shots. It had a decided wallop both fore and aft with the regulation 8mm ammunition which was picked up here and there. The initial velocity of this stuff was, I understand, about three thousand feet per second, as against the 2440 of ours. In one way, this worked to our advantage, for where the lines were close together we had a defiladed space or 'safety zone,' behind our parapets, much greater than the enemy had.

However, service issues were usually restricted to 8mm-calibre magazine rifles even though some had delicate set-triggers which reacted badly to the mud of the trenches. In addition, most were deemed too weak to handle the 'S' cartridge safely.

To prevent accidents, a tin plate fixed to the left side of their butts bore a silhouetted Patrone 88, with its distinctive round-nosed bullet, and NUR FÜR PATRONE 88 KEINE S MUNITION VERWENDEN ('only for Patrone 88, unsuitable for S Munition').

The long-term solution was clearly to modify the Gew. 98, and so, in December 1914, the Gewehr-Prüfungs-Kommission ordered fifteen thousand rifles – known as Zielfernrohrgewehre 98 or, later, as Scharfschützengewehre 98. A 1915 document describing the use of sniper rifles noted:

> With the manufacturing of the K bullet being difficult and expensive, this cartridge must be used only for precision shooting when a great

> penetration is sought. The S.m.K. cartridge is distributed only to marksmen supplied with the Model 98 rifle and telescopic sight. These men must use these cartridges wisely and not give them away.
>
> If necessary, these cartridges can be fired from a machine gun against a fortified emplacement, armoured shelter (pillbox) or armoured aircraft.
>
> The rifles with telescopic sight in the German Army are of two kinds:
>
> a) The standard rifle Model 98 on which a telescope has been installed. The first order (late 1914) was for 15,000 rifles.
>
> b) Hunting rifles with telescope. All of these in Germany have been requisitioned. These rifles have less strength than the rifle 98 and can fire the 88 cartridge only.
>
> The weapons with telescopic sight are very accurate up to 300 meters. They must only be issued to qualified marksmen who can assure results when firing from trench to trench, and especially at dusk or during clear nights when ordinary weapons are not satisfactory.

Hesketh-Prichard remarked in *Sniping in France* that he believed that, in 1915, the 'telescopic-sighted rifles of the German army' had been:

> served out in the ratio of six per company, and that these rifles were issued not to private soldiers who shot with them, but to N.C.O.'s who were responsible for their accuracy, and from whom the actual privates who used the rifles obtained them, handing them back at intervals for inspection.

Scharfschützengewehre 98 were made to a single pattern, but their sights varied in construction and performance. Though Goerz and Zeiss 4× sights seem to have been preferred, urgency meant that virtually any sight could be pressed into service.

Magnification ranged from 2¾× to 6×, though 4× was most common. The range drums of sniper rifles used by the Bavarian Army were originally set for 200/400/600 metres, while those used in Prussia, Saxony and Württemberg had 100–1,000-metre graduations. Windage adjustments were rarely achieved other than by employing a key or screw in the front mount to warp the sight-tube laterally, playing havoc with focus or even causing lasting damage.

The mounts were usually claws which simply slotted into the bases, where they were held by buttons, latches or finger-wheels. The back mount was often cranked sideways to give access to the magazine.

In some cases, the sight was offset to the left for the same reason. However, most were set high enough to facilitate reloading, while skeletal bases allowed open sights to be used if necessary. Raising the optical sight above the breech, of course, compromised the ability of the sniper to conceal himself.

British and German optical sights were usually conventional, but most of the Ross rifles of the Canadian Expeditionary Force snipers used the Warner & Swasey sight that had been accepted by the US Army prior to the war. Though an evolutionary dead end, soon replaced or supplemented by the much more conventional Winchester A5, a fragile unit with problems of its own, the prismatic 5.2× M1913 Warner & Swasey sight improved the marksmanship of those who had mastered its quirks. However, in 1917, Lieutenant William Curtis, then recently transferred from 187th Battalion CEF to serve as sniping officer of 21st Reserve Battalion, reported that though 'the lenses of the Warner & Swasey sight are good and give a good field and clear vision':

> The method of elevating the sights allows for a certain amount of play, and when a shot is fired the sight moves and does not come back to its proper place; this makes it practically impossible for two shots to be fired at exactly the same point. For reasons which no-one knows, the lateral adjustment is hard to depend upon as, when the sight is taken off the rifle, it is necessary, even under exactly similar conditions to re-adjust the sight laterally. Rubber bands have been used to help hold down the sight to its true position, but even with this it is not to be depended upon. We have in use one Aldis sight which gives good results. Men returning from a sniping course, however, are convinced that the Winchester sight is the best for practical work.

Herbert McBride, in *A Rifleman Went to War*, was more charitable:

> Now, I have heard a lot of unfavourable comments against this Warner & Swasey sight, in fact, I cannot recall ever hearing a good word spoken for it. However, it is my opinion, that when compared to the others we had at that date and time, it was a pretty good sight. Naturally, it might not compare with the scope sights of today, as much progress has been made in these since the war. The one I used gave very good results, and was fully as accurate and reliable as the Winchester A-5 type. This latter model was particularly hard to keep 'lined up'.

When the First World War ended, the Germans still relied on the SG 98 while the British had optically sighted SMLEs pending the introduction of the

Pattern 1914 Mark I* W (T). French *tirailleurs* had small numbers of specially selected Lebel and Berthier rifles fitted with French-made APX sights, though comparatively little is known about their use. Said McBride:

> As we never were associated with any French troops, I know nothing about their use of the rifle beyond what I saw one morning at the capture of Combles, when I was temporarily attached to a unit of the Gloucester Regiment, and that was at a distance of some three or four hundred yards. I have fired the Lebel on a number of occasions but simply at some mark just for fun, you might say.

The Austro-Hungarians fielded straight-pull Mannlicher Repetiergewehre M1895 and associated Stutzen, short rifles, fitted with commercial-type optical sights obtained from manufacturers who also often supplied the German Army. The Italians had specially selected Mo. 1891 (Mannlicher–Carcano) rifles with optical sights, some of which seem to have been captured from the Austro-Hungarians.

Rifles for Snipers, 1939–45

When the Second World War began, sniping had fallen so greatly from favour that virtually none of the protagonists – the USSR excepted – provided specialist equipment. Having discarded virtually all their sniper rifles and optical sights in the early 1920s, the British found themselves in a difficult position, as Clifford Shore remarked in *With British Snipers to the Reich*:

> The official sniper's rifle of the British Army [in 1939] was the P 18 – the P 14 fitted with the P 18 sight, but I cannot gather any information as to the whereabouts of these rifles. Perhaps they were in Battalions' stores, and maybe no one knew how to use them! I have seen it recorded that one of the Army's finest shots could not find a single, solitary British telescope sighted rifle in the Middle East in the summer of 1941.

Consequently, virtually anything which could shoot was impressed pending rehabilitation of the P/14 Mark I*W (T) rifles that could be retrieved. In 1940, Alex Martin & Company of Glasgow supplied 421 'No. 3 Mk I* (T) A' rifles with wooden cheek pieces and Aldis-type P/18 telescope sights in low mounts offset to the left of the bore. In addition, good use of No. 3 rifles was made by Australian snipers in North Africa and the Far East, supplementing a heavy-barrelled version of the SMLE, and by Canadians pending the availability of the No. 4 Mk I* (T).

Though approved on 15 November 1939, no issues of the No. 4 Mk I (officially redesignated 'No. 4 Mk 1' in September 1944) were made until the spring of 1942, though more than two million had been made in Britain when the war ended. In addition, Small Arms Limited of Long Branch in Canada and the Savage Arms Company of Chicopee Falls in the USA made 2.1 million of the No. 4 Mk I* rifle approved in June 1941.

Clifford Shore, writing in *With British Snipers to the Reich*, was initially unimpressed:

> Many of the early rifles were poor and had many faults, at one stage opinion was dead against the model. Bolts needed a lot of work before they functioned decently; rear sights had considerable lateral play; magazines were faulty; the bores were poorly finished due to lack of machine operations, and there was great tolerance and allowance manifest when passing the bores for gauging ... Some of the rifles were terrible.

However, by 1942, Shore admitted that 'things were looking up' and that:

> An establishment of eight snipers per Battalion had been approved, and something had been done, or at least started, towards equipping those snipers with a good rifle fitted with telescopic sight.

The first attempt to provide a new Lee–Enfield sniper rifle was made in May–June 1940, when 1,403 No. 1 Mk VI and No. 4 Mk I trials rifles were converted at Enfield. The perfected No. 4 Mk I (T) was then approved on 12 February 1942, accompanied by the No. 32 Mk I telescope sight. Rifle and sight were numbered to each other to ensure that performance could be restored if the components were ever parted. Beginning in September 1942, London gunmakers Holland & Holland delivered the first of the 23,177 No. 4 Mk I (T) rifles made up to August 1945.

Most of the No. 4 Mk I* (T) rifles assembled in Canada were fitted with a variant of the No. 32 or, more rarely, had an American-made Weaver sight in a Griffin & Howe rail-type mount attached to the left side of the body.

No. 4 Mk I (T) rifles were selected from those that had shown excellent accuracy during acceptance tests. Their No. 32 sights were fitted individually, pairings being accepted only if they could place seven shots of seven in a four-inch circle at 200 yards, and six out of seven in a ten-inch circle at 400 yards.

Individual rifles were often capable of a minute-of-angle performance, which would mean a four-inch group at 400 yards. Consequently, many recipients were impressed. Denis Edwards noted in his diary on 25 July 1944, reported in *The Devil's Own Luck*, that to his 'surprise and delight' he:

> was issued with a brand new sniper rifle [No. 4 Mk I (T)] straight out from Ordnance. It was covered with a thick layer of grease and wrapped in greaseproof paper. I spent much of the day taking it apart and cleaning it. Until this time my rifle had been a completely standard issue Lee Enfield .303 Mk. IV [No. 4 Mk I], with no telescopic sights, and this new weapon was much more suitable.

But the process had been compromised at the outset, because, as Clifford Shore observed, sniper rifles were often in the wrong hands:

> In one company the new company commander, a pre-war rifleman, starting a blitz in the sniping department (there were two snipers in each company) found that neither of the then 'snipers' had had any special training, and that the rifles were not zeroed, and were virtually useless. After asking for volunteers he picked out three good shots who were very eager to have a crack at becoming snipers, whisked them out of the line and with the aid of an armourer got the rifles in good condition again and zeroed into the men who were going to have a shot at the Huns.

Almost alone in Europe after the First World War had ended, the Russians could still see the value of snipers and their specialist equipment. In *Red Army Sniper*, Evgeniy Nikolaev remarked that the:

> ability to hit an enemy with a single shot is acquired through determined study and practice, thorough knowledge of one's weapon, and meticulous care of it. Our own home-manufactured rifle, the 1891/1930, a first-class weapon in all respects, designed by the talented Russian engineer Sergei Mosin and equipped with telescopic sights, has deservedly won the love and acknowledgement of all Soviet snipers.

The Obr. 91/30g, approved in April 1930, was made in Tula and Izhevsk in amazingly large quantities – including about 54,160 sniper rifles set up in Tula prior to 1938/9, when work stopped in favour of the Tokarev, and at least 275,000 made in Izhevsk in 1942–58. Tula also contributed small quantities in 1943–5. Most pre-1945 examples displayed C and П above the chamber, *Snayper proveryat* or 'Sniper[-issue] approved'.

Output exceeding 250,000 in 1941–5 alone makes the Mosin-Nagant numerically, if not necessarily technically, the most influential sniper rifle of the Second World War. It also ensured widespread distribution. Virtually every published image of Vasiliy Zaytsev and Roza Shanina, for example, shows one.

Individual snipers often tried to improve their otherwise-standard-issue weapons. Lyudmila Pavlichenko, whose memoir *Lady Death* is one of the few to feature weapons in detail, noted that to bring 'my new sniper rifle into battle condition' she had had to 'strip the rifle completely':

> and do some work on its components. For example, I had removed the wood along the whole length of the handguard groove, so that the woodwork no longer touched the barrel, I had filed down the tip of the gunstock, so that the barrel fitted snugly, bedded the barrel properly in the fore-end, and inserted padding between the receiver and the magazine. To ensure that the various parts of the bolt mechanism worked properly, it was recommended that they were carefully worked over with a small needle file. The trigger mechanism of the rifle needs to be efficient, reliable and stable.

She added that it was essential to test the rifle after changes had been made, however insignificant they appeared to be:

> The clear and windless weather which had settled over the steppes of the Dnestr's east bank was conducive to adjusting rifles, first with open sights and then the telescope. The first stage was undertaken at a distance of 100 metres from a square target measuring 25 × 35cm. For the second stage the rifle was fixed on a stand (or something similar) to ensure its immobility. The objective was to regulate the telescope sights with the aid of the open sights.

The first Mosin-Nagants were fitted with the 4× PE (Obr. 1931g) sight, the culmination of experiments begun in the late 1920s, which reached service in 1932. Long and heavy, with a focusing collar ahead of the ocular lens, the PE had separate elevation and windage adjusters.

Rifles tested in the Spanish Civil War (1936–9) passed scrutiny, but recoil could jolt the annular collar of the PE sight out of focus and ineffectual seals promoted excessive fogging. The improved PEM (Obr. 31/37g or Obr. 1937g) was essentially a PE without the focus adjuster. The design of the sight base changed to accommodate a cylindrical fore-body, while the new side-rail was retained by a peg-and-slot connector and a large-headed bolt.

The short-bodied PU sight had been introduced in 1940 for the Tokarev and then, after proving reliable, was adapted to fit the Mosin-Nagant. A new mount was required, so a baseplate was attached to the left side of the action-body by pegs, screws and set-screws. The two-ring monoblock attached to the baseplate rail by a ball-and-socket at the front and a large knurled-head screw angled against the base at the rear. Set-screws in lugs at the rear of the baseplate allowed elevation adjustments to be made during setting-up.

Though the Obr. 91/30g was popular with most of its users, including several well-known Germans as well as most Russian marksmen, not everyone appreciated it. Simo Häyhä, for example, rejected a captured Obr. 91/30 fitted with a PE sight because he believed he would have to lift his head to look through the telescope and, therefore, that light glinting from the objective lens would betray his position. Häyhä's memoirs confirm his personal rifle to have been Mosin-Nagant m/28-30 Sk.Y. no. 60974, the standard pattern issued to the Protective Corps prior to the introduction of the perfected universal-issue m/39.

When the Germans invaded the USSR in June 1941, most Soviet snipers carried selected 'three-line' rifles fitted with PE or PEM optical sights even though the Soviet authorities believed the auto-loading Tokarev SVT to be an improvement on the manually operated Mosin-Nagant.

The Snayperskaya Vintovka Tokareva (SNT) had superseded the Obr. 91/30g sniper rifle early in October 1940, nearly 50,000 being made in 1941–2. Adapting the Tokarev for sniper use, however, had posed problems as the gun ejects up and to the right. The answer was found in a short-body 3.5× sight, the Pritsel Ukorochenniy (PU), which had been created in the Feliks Dzherzhinsky manufactory operated in Kharkov by NKVD.

Approved on 18 July 1940 for issue to NKVD units, the PU was carried in a cantilever mount with arms that extended forward over the receiver but stopped short of the ejection port. Experience showed the sights to be too flimsy and unnecessarily difficult to make, leading to a modified version with the adjuster drums mounted directly on a steel tube-body.

Smaller, lighter and easier to make than the PEM, the PU performed well enough in service even though the absence of a focus ring meant that eye-relief was critical and only firers with good eyesight could make the best of it. In addition, the objective lens, without any protective shade or notable recessing, was prone to glint in sunlight and reveal the firer's position.

However, the cantilever sight-mount issued with the SNT proved to be insufficiently rigid to maintain zero. Combined with inaccuracy beyond

800 metres compared with the Mosin-Nagant, this was sufficient to ruin the Tokarev's reputation even though trials have since shown the rifle to be acceptably accurate and reliable as long as good-quality ammunition is used.

Yet snipers Nikolay Aksakov, Fyodor Chegodaev, Nikolay Ilin, Fyodor Okhlokpov, Lyudmila Pavlichenko, Vladimir Pchelintsev and Ivan Sidorenko were among those who put Tokarevs to good use. Aleksey Grinberg, contributing 'Odin iz geroicheskoy kogorty' ('One of a Heroic Cohort') to the Nikolaevs' *Sovetskie Snaypery v Boyo*, related how Chegodaev had taken receipt of SNT no. 441 [prefix letters unknown]:

> Beneath its armoury grease, it seemed to have preserved the warmth of its previous owner, just as old paintings retain their bright colours under a layer of age-old dust. 'Well, you're going to have to serve me now, old girl', said Chegodaev, 'and I'll look after you'.
>
> On his first day in the new area Chegodaev wiped out eight Nazis and two horses. He lay on the roof of an old tumbledown barn and fired methodically, one shot after another. The following day Chegodaev prudently selected a different position; a sniper must not stick to one site ... The Nazis had only discovered the sniper by accident after he had fired ten shots from the roof of the barn.
>
> Chegodaev now operated more thoughtfully and cautiously; he took three or four shots and hid. The enemy would spot the sniper's position from the flash of the muzzle, the puff of snow dust and the sound of the shot, and try to annihilate him. After three shots Chegodaev crawled away to a different site. And no sooner had he set himself up again than machine guns and mortar guns opened fire on his old position. Chegodaev noticed that their firing was very accurate. 'Blow the whole world away if you like,' he scoffed ...
>
> He operated calmly, cunningly, and after careful consideration ... The tally of dead Nazis he kept was already into three digits – one hundred, two hundred, three hundred, four hundred, five hundred. Self-loading rifle no. 441 served him faultlessly.

In *Lady Death*, however, Lyudmila Pavlichenko commented that the Tokarev:

> would begin to 'snap' (for instance, it would not reload or ... barely eject spent cartridges) if the pressure of the powder gases changed. It depended, incidentally, on the weather and even on the air temperature. In these cases the marksman had to regulate the aperture in the gas port manually ... Apart from that, the 'Sveta' also

> misbehaved when covered with thick grease and if dust got into its mechanism.
>
> Among the deficiencies of the SVT-40 I would also mention its bright muzzle flash upon firing (on account of a barrel that was 100 mm. shorter than the 'Three-line') and its loud sound, which immediately gave one's location away.

But Pavlichenko recognised that the auto-loading Tokarev had advantages in specific circumstances:

> It was superbly suited to clashes with the enemy in the field, when artillery, machine guns, and mortars were operating. However, to put it bluntly, it increased the danger of a marksman being detected by the enemy in a single-person hideout, for example, in a forest. However, among the sniper fraternity, the 'Sveta' had its admirers.

Though the Red Army showed an ambivalent attitude to the SVT, deemed much too complicated to issue to ill-educated infantrymen, the Finns were only too pleased to turn Tokarevs captured during the Winter War against their erstwhile owners. And Germans on the Russian Front, particularly in the early part of the war, were also keen to acquire rifles with better firepower than the Kar. 98k and longer range than submachine guns.

In 1941, however, the destruction of huge tracts of the western USSR forced the wholesale relocation of industrial plant. Consequently, work soon concentrated on the Mosin-Nagant: simpler, more robust, easier to maintain and less prone to jam. Production of Obr. 91/30g sniper rifles recommenced in March 1942, and the SNT was officially withdrawn in October 1942. By the time the sniper schools were created, therefore, the Red Army's flirtation with the auto-loading Tokarev had virtually ended.

The standard German sniper rifle was the optically sighted Zielfernrohr-Karabiner 98k (Zf.-Kar. 98k) – though notable shortages brought surviving full-length *Scharfschützengewehre* out of storage when the war began. Zielvier, with 4× magnification, was regarded as standard, but individual sights were simply acquired on the commercial market, and, therefore, came in a profusion of shapes and sizes. Sights were customarily clamped in a side-rail mount held to the left side of the body. High mounts were often cut away to allow open sights to be used, but, as they forced the sniper to reveal his head, were a liability in combat.

Unfortunately, the considerable power of the 7.9mm rifle cartridge rapidly loosened the screws holding the rail, and many remedies had to be tried.

These included post-and-claw mounts, acquired from commercial sources, and turrets made in 'high' and 'low' form depending on the diameter of the ocular-lens tube. But none proved effective, as the comparatively short rail was insufficiently rigid.

The invasion of Russia in the summer of 1941 exposed the German Army to snipers who exacted a dreadful toll. The Obr. 91/30 sniper rifle proved to be more accurate than the Zf.-Kar. 98k, which even the Germans admitted, and the SNT (Tokarev), while incapable of matching the bolt-action weapons for accuracy, offered a greater rate of fire.

Captured rifles armed not only Wehrmacht infantrymen but also snipers: Sepp Allerberger achieved his earliest kills with an Obr. 91/30 retrieved from an arms-dump, and Oberfeldwebel Hellmut Bommel, the ill-fated opponent of Lyudmila Pavlichenko in one of the best-known duels, was found to have been using an Obr. 91/30 with a PE sight.

The Wehrmacht issued large numbers of *Zielfernrohrgewehre* early in the war, but this term simply means 'telescope-[sighted] rifle' and does not refer specifically to snipers: the goal had simply been to improve the performance of infantrymen until 'everyone was a marksman'. Thus the hopelessly ineffectual 1.5× Zf. 41 was a sighting aid masquerading as a sniper sight – impressing almost no one whose life depended on it, even though an adaptor allowing the Zf. 41 to be fitted to the back-sight bed of any Kar. 98k was developed.

The firepower of Tokarev auto-loaders soon forced the German authorities to respond. Differing ideas had soon been submitted by Mauser and Walther, becoming the Gewehr 41 (M) and Gewehr 41 (W) respectively, but field trials then showed them to be unreliable. However, the Walther version was deemed to have promise.

Duly adopted in December 1942, about 123,000 were made, including a few Zf.-Gew. 41s with Zielvier offset to the left of the breech to allow ejection to take place efficiently. However, the muzzle-cone mechanism, susceptible to corrosion, was then replaced by one inspired by the Tokarev.

A detachable box magazine was substituted for the fixed charger-loaded case, the barrel was shortened, and the Gewehr 43 was adopted in April 1943. No sooner had series production begun than critical raw materials began to run short, machine-time had to be selectively allocated to maximise production, and the incorporation of castings and stamped parts promoted the breakages and extraction problems that dogged what could have been the best semi-automatic rifle of the Second World War.

The Gewehr 43 Zf. had a new 4× sight, originally developed for the FG. 42 but inspired by the Soviet PU, in a special one-piece mount clamped to a rail on the left side of the body. However, though more than 50,000 Walther sniper rifles were made, they were generally unpopular.

Asked in 1969 what weapons they had used, Matthäus Hetzenauer noted that he had had a Zf.-Kar. 98k with 6× telescope sight and then a Zf.-Gew. 43 with a 4× sight; Sepp Allerberger had begun his exploits with a captured Russian sniper rifle, exchanging it for a Zf.-Kar. 98k with a 6× sight; Helmut Wirnsberger had been issued a Zf.-Kar. 98k with the ineffectual 1.5× sight, then a similar rifle with a 4× sight and, lastly, a Zf.-Gew. 43 with a 4× sight.

Matthäus Hetzenauer said that he considered the standard 4× sight to be sufficient at ranges up to about 400 metres, but a 6× sight was good up to 1,000 metres. Sepp Allerberger, whose captured Russian 91/30 rifle had a 4× sight, opined that a 6× sight mounted on a Zf.-Kar. 98 was ideal. And Helmut Wirnsberger, after dismissing the 1.5× Zf. 41 as ineffectual, agreed that the 4× sight had proved to be good.

Hetzenauer and Allerberger suggested that there was no need for a sight rated above 6×, though neither had had experience of greater magnification. Wirnsberger regarded 4× as sufficient for all purposes. Interestingly, Herbert McBride, writing of the First World War, broadly concurred. In *A Rifleman Went to War*, he had noted that:

> For the general purpose telescope, to be used in all sorts of weather and conditions – and by advancing troops – I believe the three power to be about as high as one can go, possibly the two-and-one-half power is enough, at that. Large field and clear vision is more important than magnification in most instances. But when it comes to real honest-to-God sniping, from a prepared position and with an observer to help, where the ranges may run up to seven and even eight hundred yards, one can use considerably more magnification, and even an eight power telescope is none too high.

Replying to questions of preference, Hetzenauer stated that 'of all weapons available at that time it [the Zf.-Kar. 98k] had the highest accuracy for permanent use, besides it did not jam easily. G43 was only suitable to about 400 metres. It also had inferior precision.'

Sepp Allerberger concurred: 'K98 was best. The G43 was too heavy', but Helmut Wirnsberger thought the 'G43 would have been good if it did not jam easily and its capacity [accuracy?] was as good as K98'. Declining manufacturing standards were not compatible with extreme accuracy, postwar trials revealing that the Zf.-Gew. 43 was considerably inferior to the Zf.-Kar. 98 in this respect.

However, asked how they would choose between the Zf.-Kar. 98 and a 'semi-automatic rifle that does not easily jam and has the same [magazine] capacity', Hetzenauer responded that 'Snipers do not need a semi-automatic weapon if they are correctly used as snipers', while Allerberger preferred a 'semi-automatic loader, if its weight does not increase', and Wirnsberger concurred on the grounds that 'Faster firing [was] possible when attacked by the enemy'.

Adoption of the M1903A3 Springfield for the US Army in May 1942 allowed development of a sniper rifle to proceed, trials undertaken by the Infantry Board and the Ordnance Department leading to approval on 14 January 1943 of the 'Rifle, Caliber .30, M1903A4 (Snipers)'. The Model 330C optical sight made by Weaver of El Paso and the 'Junior' mount made by Redfield of Denver were standardised at the same time.

Remington had suggested at the outset that snipers would have benefited from using a militarised variant of the Model 720 bolt-action sporting rifle, but the Ordnance Department was wary of losing the virtues of commonality of parts that lay in adapting the Springfield.

Ironically, therefore, a 20,000-gun contract for the M1903A4 was duly passed to Remington, followed by another order on 20 June 1943. It is no longer clear precisely how many had been made by the time the last batch was accepted in June 1944, but the total neared 30,000. Numbered from 3407088 to 3427087, guns were taken from M1903A3 production, selected for accuracy, and honed to ensure they cycled smoothly. The bolt-handle shank was given an 'outcurve', allowing it to lie closer to the stock and ensure the bolt handle cleared the sight-tube during the loading stroke.

Many soldiers saw the M1903A4, first used in combat during Operation Torch, landings made in North Africa in October 1942, as a poor design. The optical sight was fragile, often breaking the adjuster drums, and ineffectual seals led to fogged lenses, encouraging the growth of mould in the tropics. Small-diameter lenses gave a narrow field of view and comparatively poor light-gathering abilities, while individual rifles often 'lost zero'. The omission of open sights meant that firers experienced difficulty when engaging targets rapidly at short range or if the optical sight could not be used.

The US Army had always considered replacing the Weaver 330C sight and its M73B1 militarised variant with the M73 Lyman Alaskan pattern. However, very few Lyman sights had been fitted to M1903A4 rifles by the end of the war and it is unclear if any saw combat.

Marine Corps scout-snipers who landed on Guadalcanal in August 1942 carried M1903 Springfields equipped with Winchester A5 or Lyman 5A scopes. However, the Springfield had not been designed specifically for long-range accuracy, with a comparatively short, thin-walled barrel, and some marine commanders viewed it merely as a stopgap.

USMC Colonel George Van Orden was not only a proponent of a cohesive sniper-training programme, but had repeatedly demanded the introduction of a dedicated scope-equipped sniping rifle that was not just a modified service weapon. Even though his views were regarded with suspicion by his superiors, Van Orden's logic was difficult to fault.

In the hands of US Marines Corps snipers, therefore, the M1903A4 was sometimes replaced by the Winchester Model 70 sporting rifle, the Weaver sight being abandoned in favour of the high-magnification (if somewhat delicate) Unertl target patterns. Featuring a two-lug Mauser-type bolt with a patented bolt-guide rib to smooth the operating stroke, 373 Model 70 rifles had been acquired by the USMC from Winchester in May 1942: numbered in the 41000–50000 range, they were supposedly destined for training.

The adoption and widespread issue of the auto-loading Garand was a game-changer. Teething troubles had been largely overcome by 1941 and mass production in Springfield Armory was under way even though shortages forced the USMC to procure the recoil-operated Johnson – designed by an ex-marine – even though it was much more delicate than the Garand.

Yet there is evidence that a few Johnsons, including some Netherlands East Indies Army guns retrieved by the Australians, were fitted with optical sights to serve snipers in the absence of anything better.

The M1C Garand sniper rifle, standardised in July 1944, was fitted with an M73 (Lyman Alaskan) or an M73B1 (militarised Weaver 330) optical sight in a Griffin & Howe mount offset to the left so that an empty clip would be ejected satisfactorily and the magazine could be reloaded. The M1D version was introduced in September 1944, but few had been made when fighting in the Pacific ceased in August 1945.

The firepower of the Garand often proved to be an advantage, particularly when multiple targets had to be engaged in snap-shooting. Many soldiers who showed marksmanship abilities made good use of it. Private Carl Cartledge of the intelligence and reconnaissance team attached to the 101st Airborne Division, the 'Screaming Eagles', eulogised the performance of his colleague Roland Wilbur, whose:

> legendary skill with an M-1 languishes in obscurity. It should not. Gifted with extraordinary eyesight, each round that Wilbur fired served as practice to improve his skill. Wilbur fired tens of thousands of rounds to push himself to the limit. After Normandy a man … watched with powerful binoculars what Roland Wilbur did with the

> naked eye. He stated 'this man can shoot his rifle further and faster than our calculations say it can be done'.
>
> Wilbur could snap shoot between five and seven hundred yards. He later recorded kills at that distance at Bastogne. Wilbur never fired from the same place twice. He never exposed himself longer than to roll over. If no-one lined up in his sights, Wilbur disappeared. When Wilbur fired, there was one less of them.

The major drawback of the Garand was the clip, which automatically ejected after the last round had been fired. This prevented snipers selecting cartridges for specific tasks, unless the clip had been pre-loaded, while the optical sights had to be offset so that the clip missed them during ejection. In addition, a spent clip could make enough of a noise while landing to betray the sniper's position.

While some of their American opponents had auto-loaders, Japanese snipers used 6.5mm Model 97 or 7.7mm 99 Arisaka rifles, customarily fitted with optical sights, but they retained open sights largely because long-distance shooting was rarely appropriate in the jungle.

It was argued that in such conditions well-camouflaged marksmen could shoot accurately to 400 yards, without wasting the time sometimes lost while acquiring the target through the telescope. However, the Japanese were taught to shoot to 600 yards in a variety of conditions, so that each man would understand how the point of impact varied with wind, temperature and many other factors.

Such preparation could be judged sketchy by Allied standards, but was adequate for much of the fighting undertaken in the Far East. But many Japanese sights, though often excellent optically, lacked the adjustments that Western marksmen would have considered standard features. This simplicity removed the need to re-zero sights in the hands of inexperienced firers, but depended greatly on the skill of the individual sniper to estimate aim-point.

Snipers were encouraged to fire from a rest wherever possible, a folding-wire bipod being fitted to the rifle's fore-end. However, though intended to steady aim, flimsiness proved to be its weakness. Some men carried a small armour plate with an earth-spike protruding downward and a 'U'-cut on its upper edge to receive the fore-end.

Though the 7.7mm cartridge had been adopted in 1939, after combat experience in China, many of the Japanese snipers engaged in the Pacific theatre, according to the *Marine Corps Gazette* of January 1945:

> still use .25 weapons. The .25 calibre weapon will not cut brush or penetrate like a .30 calibre weapon. It will however, penetrate our

helmets at ranges of 150 yards or more. Its most annoying characteristic is that for practical purposes the powder is actually smokeless and generates little muzzle blast. Thus a sniper can 'hole-up' or 'tree-up' and although he may fire considerably, we will seldom, if ever locate him by smoke or muzzle blast.

Testing a captured 6.5mm Arisaka rifle had impressed US Army Lieutenant John George:

> I fired at the short ranges – as near 100 metres as I could estimate. All the shots were bulls. At 200 metres the weapon kept everything inside 8 inches, even with some bad holding involved. The gun continued to behave beautifully, not missing any of the settings for ranges up to 400 metres.
>
> To check the zero-return qualities of the scope mount, I took the scope off the rifle several times during the firing, with no noticeable change in the point of impact.
>
> There was no ready adjustment on the scope but at least the Jap soldier would not be able to ball-up what zero his rifle had been given at the arsenal.

Trials and Tribulations

If problems arising with the rifles were bad, those associated with telescope sights could be far worse. Hesketh-Prichard gave a good illustration of their scale in *Sniping in France*.

After discussing the way in which optical sights were issued in the German Army in the early stages of the First World War – to experienced NCOs who ensured that the rifle/sight combinations were properly regulated before handing them to individual snipers – he observed that this method was better than on the British side, where, all too often, the man using the optical sight knew nothing about it:

> On one occasion I had gone down on duty to a certain stretch of trench and there found a puzzled-looking private with a beautiful new rifle fitted with an Evans telescopic sight. 'That is a nice sight,' said I. 'Yessir.' [The private replied] I examined the elevating drum, and saw that it was set for one hundred yards. 'Look here,' I said 'you have got the sight set for a hundred. The Hun trenches are four hundred yards

away.' The private looked puzzled. 'Have you ever shot with that rifle?' I asked. 'No, sir.'

And Herbert McBride observed in *A Rifleman Went to War* that:

> Whether or not the sniper can get away with the job of handling a rifle equipped with a telescope sight in battle is a question for the future. I have never heard that it had been successfully accomplished. The fact is that the rifleman has to squirm and crawl through every little hole he can find; often through deep mud and water, over rocks and through brush. It is a sure thing that he will have 'one hell of a job' to keep that scope from being broken or so badly disarranged as to make it useless.
>
> Of course, there have been and will be battles fought over ground where the conditions would make it comparatively easy to 'baby along' a telescope, but we cannot depend on that – it will be the exception and not the rule. The scope would certainly be a great help in picking out the obscure targets – if it could be taken along.
>
> Since the war I have read a great deal about the new types of telescope sights and their improved mountings. Many of our leading sportsmen and big-game hunters are claiming they are just as substantial and dependable as the iron sights. I cannot by any means subscribe to this contention, because my personal experience, gained in the trenches and in battle, is all to the contrary.

In addition, McBride said, 'targets are mighty hard to locate on the battlefield', and also:

> Under such conditions as these, the rifle must be fitted up so the magazine is available for use at all times; no modern rifle-shot, no matter how good, is going to advance very far with a single shot rifle in his hands.
>
> My own ideas as to the proper telescope for such battle firing would be one with an absolutely rigid mount that would stand up under any and all conditions of jar and abuse. For that matter, the rigid mount is the only type to consider for sniping work of any description – one could not begin to use these conventional target mounts where the scope slides forward from recoil and must be pulled back by hand after each shot. The whole works want to be solid and rigid, with no outside springs, shock absorbers, or similar gadgets in sight. If these must be

> used, put them inside of the scope where they will be out of the way – and out of the dirt and dust also.

Whether a telescope sight should remain on the rifle as long as it was being used, or detached as soon as possible into a protective case, was hotly debated. The Red Army was the greatest proponent of sights remaining on the guns at all times, possibly because its snipers, as a group (generally excepting women), were uneducated; the Germans were most likely to constantly mount and remove their sights, with unfortunate effects.

Experienced snipers would always fire warm-up and zero-check shots when they could, but mounting a telescope sight under pressure, to fire a quick shot, has always been risky.

Personal-defence and Other Weapons

Snipers were at great risk each time they took to the field, gambling on their rifles and optical sights performing properly. If their positions were overrun, bolt-action rifles could not be reloaded quickly enough to repel mass-attacks. One obvious answer was to carry back-up weapons.

In addition to knives, handguns were readily available: regulation issue, bought privately or taken from the enemy. The British and associated forces relied largely on six-shot Webley and Enfield revolvers during the Second World War, but most other armies carried effectual self-loaders such as the German Luger P. 08 and the Walther P. 38, the Soviet Tokarev (though large numbers of Nagant revolvers remained in service) and the M1911A1 Colt–Browning of the US forces. In addition, huge numbers of sophisticated small-calibre blowbacks such as the German Walther PPK and PP, the Mauser HSc and the Sauer 38H – often taken from slain officers and downed aircrew – were used as back-up weapons by snipers.

Clifford Shore drew attention to the 'many Englishmen' who 'appear to have a great distrust of all handguns', adding that it was amusing:

> to see how some men handled a pistol or revolver for the first time; gingerly, as though the barrel were made of some plastic substance which would playfully curl up and come to rest in their direction when fired. It may appear a rather curious innovation, but I maintained that it was part of a sniper's training to become familiar with the handgun – one never knows when such a facility will be needed – so familiar indeed as to be able to draw with speed, and

> shoot with either hand with a certainty of hitting a man at ten or twelve yards, not once but two or three times in the most brief space of time.

But that when the war had started:

> British factories could not turn out the Enfield fast enough, so the Smith and Wesson people were asked to augment the supply of .38 pistols. I used a Smith and Wesson pistol for nearly four years and I have nothing but praise for it …
>
> So much has been written about the Luger and the Walther 9mm pistols that I consider it unnecessary to add much to the ever-growing volume … I found the hang, balance and grip of the Walther to be better than the Luger, and have nothing but admiration for the craftsmanship of these German pistols, but I am still a sworn adherent to the standards of the Enfield and Smith and Wesson for a sturdy, reliable combat job.

Lyudmila Pavlichenko was particularly fond of her personal sidearm, a 7.62mm Tula-Tokarev or 'TT' pistol. Promotion to senior sergeant had entitled her to wear 'a leather belt with a shoulder strap, a single-pin brass buckle and a holster with ramrod to accommodate a pistol':

> I never subsequently parted with this firearm, the Tula-Tokarev given to me in Sevastopol. I had it with me in sniper's hideouts, on leave in the city and, of course, on parade, and, following evacuation, it accompanied me to Novorossiysk and then to Moscow.
>
> The TT became my talisman. While hunting for Fritzes in the Crimean forest I relied, in the event of failure, not so much on the grenade which always hung from my belt, but on my 'Totosha' (as we called the Tokarev pistol in the army).
>
> Neither the Russians nor the Germans took snipers prisoners, but shot them directly on the spot. For women there was another variation – death preceded by gang rape. Therefore, the grenade was for rolling under the enemy's feet, seven bullets from the TT were for anyone who came too close, and the eighth was for yourself.
>
> I do not deny that a pistol weighing 825g without its eight-cartridge magazine is somewhat heavy for a woman's hand … The most important thing was that the TT met all front-line demands: a powerful 7.62mm-calibre cartridge which could penetrate a brick

> wall 100mm thick, a sturdy barrel, a reliable trigger mechanism and a conveniently shaped butt.

Another answer was found in the submachine gun, whose large-capacity magazine allowed fire to be sustained for much longer than even the auto-loading handguns. The Germans made good use of the Maschinenpistole MP38 right from the start of the Second World War, followed by the MP40, but the British and the Russians were poorly equipped until the mass production began of the Sten Gun and the PPSh respectively. The Australians procured the Owen, the US armed forces used versions of the Thompson before settling on the M3 Grease Gun, but the Japanese Army, and by extension its snipers, had nothing comparable.

Snipers were often forced to rely on submachine guns for self-protection and while involved in assaulting enemy lines. Sepp Allerberger always sought to be in the forefront of action, explaining in *Sniper on the Eastern Front* that:

> Often the distance between himself and the enemy dwindled so fast that he had to put down his sniper's gun after a few aimed shots and take up the MP40 which he always had slung ready on his back during such situations. Fighting of this sort was always very problematic for Sepp, because the front lines rapidly dissolved into close-quarter combat, and at distances below 30m the telescopic sight of his gun was virtually useless because its range of vision narrowed too quickly.
>
> Aiming below the telescopic sight through the open sight was not practicable either, because the telescopic sight restricted his range of vision to a very narrow slit. Consequently such situations placed a marksman under a disproportionate amount of stress, since he could not afford to lose his telescopic rifle and yet being seen with it exposed him to the special danger of being singled out by the enemy. Identified marksmen [snipers] were invariably subjected to especially intense fire.

And Lyudmila Pavlichenko, referring to 1942, said in *Lady Death* that:

> According to the establishment laid down by the People's Commissariat of Defence on 5 April 1941, a rifle platoon was composed of quite a large group – fifty-one troops. It was commanded by a lieutenant armed with a pistol; he had a deputy with the rank of senior sergeant, who had a PPD-40 submachine gun and a dispatch rider (for communication with superiors) with a Mosin rifle. The platoon

> consisted of four rifle sections, which were headed by sergeants (all armed with SVT-40s).
>
> Attached to the platoon was a mortar section (four men along with a sergeant plus a 50mm mortar) …
>
> The Degtyarev-designed PPD-40 and, later, the Shpagin's PPSh-41 – with drum magazines for the seventy-one pistol-cartridges – were indisputably effective weapons in close combat, but there was a catastrophic shortage of them in rifle units during the first months of the war. It would appear that our two platoons of regimental scouts had only about twenty-five to thirty PPD-40s between them.

The US Army's auto-loading .30 M1 Carbine, standardised in October 1941, became a particular favourite as it was light and handy, fired a cartridge that was much more powerful than a pistol, and could strike accurately at 200 yards. Some patterns could fire automatically, and a surprising number produced towards the end of the war were fitted with infrared Sniperscopes.

Snipers frequently learned to handle many weapons, often because captured items had to be pressed into service when regulation weapons ran short. Herbert McBride, though feeling less confident 'discussing the merits of the foreign rifles', suggested that anyone interested should consult the work of 'Captain Crossman' or others who were suitably qualified, and acknowledged in *A Rifleman Went to War* that:

> During the course of any engagement, if your side is successful and you get into enemy territory, you will be sure to find rifles scattered all around, and, if fighting over ground that has been contested for a year or more, as we were during the first winter, you can pick up any number of arms which have been used by the various combatants. These things were so common that we paid no attention to them other than, perhaps, pick up this or that one and play with it – just to see what made it tick, as you might say.

Several Russian memoirs testify that *Gewehre und Maschinenpistolen*, and sometimes also accessories such as optical sights, were regularly turned against their erstwhile owners. Rem Altshuller, as detailed in Ulanov's *Stalkers of the Enemy*, recalled:

> I was issued a three-line rifle, but with German Zeiss sights, and training began. If you adjusted it well, the three-line was an excellent rifle, dependable and very simply constructed. The sight protruded

> quite a long way towards the stock and therefore the bolt handle on my rifle was bent over closer to the stock's fore-end, so that it wouldn't catch. The sight itself was better [than Soviet patterns] only in one respect – it had a rubber eye-shade.

Lyudmila Pavlichenko gave an idea of the reality that existed on the front line when regulation weapons were in short supply:

> The company now had excellent armaments. We acquired a lot of weapons in combat: rifles of various models, machine pistols (the German MP. 40, more familiar to us under the name 'Schmeisser'), Soviet TT pistols, foreign pistols such as Mausers, Berettas and Steyrs, Nagant revolvers, DP light machine guns, and a large supply of cartridges. We had learned from our bitter experience of border skirmishes and, when the weaponry was insufficient, we created our own company caches. It was onerous to drag all this from place to place, but we did not want to surrender anything to the regimental store.

In *Stalkers of the Enemy*, Boris Godov confirmed that the same happened even after Soviet forces took to the offensive and the Germans began to retreat:

> We pursued them, picking up a lot of weaponry as trophies. I acquired a Parabellum pistol and Mauser sniper rifle with sights of eight-fold magnification. Nothing like the four-fold magnification I was used to. I found it in a German cart which had got stuck in the snow and been abandoned; we captured it without a fight. The lads would collect enemy sub-machine guns and cartridges and exchange them for vodka.

In the Red Army, trainees had to master the Tokarev pistol (TT), the Degtyarev (PPD) and Shpagin (PPSh) submachine guns, the Degtyarev light machine gun (DP), the heavyweight Maxim (PM) and even large-calibre anti-tank rifles (PTRD and PTRS). Vasiliy Zaytsev remarked that Sasha Gryzaev:

> Often used an anti-tank rifle to take out enemy pillboxes or blast other fixed positions. Our anti-tank rifles were two feet taller than I am, and very heavy, weighing more than forty pounds. But the anti-tank rifle was like a toy in Sasha's big mitts, and he could tote one around all day without getting tired. In battle he knew how to select the perfect position and level shots from the anti-tank rifle directly into

the embrasures of the enemy's strong points. In his hands, it worked perfectly as a one-man portable cannon.

And Antonina Kotlyarova recalled:

> I was there as a submachine gunner, our sniper skills did not work there. Because, first of all, while you aimed, you could get killed. So we acted as submachine gunners. There were some incidents: you would be looking through your sight – aha, there would seem to be several Germans. You would throw a grenade, it would appear like it did its job. In short, after liberating Warsaw, we moved on.
>
> After we forced the Vistula, there was some hill. We, girls, and another 5–7 guys were left to hold it, and our unit moved on driving the fascists before them. So we got to hold it for two days. In the night Germans were trying to capture a 'tongue'. Had they found out who was facing them, they would've trampled us. But they didn't find out. We didn't allow them. We repulsed all attacks, I even fired an anti-tank rifle and a machine gun. It had a powerful recoil. I didn't get to fire my rifle there either, only sometimes used its sight for observation. We held on to that hill, and then our troops arrived. We could even be employed as machine gunners, as we had already learned how to fire the regular Maxim. We could do everything on our own.

Sniper on the Eastern Front describes Sepp Allerberger's role in repulsing a night-time assault, exchanging his Zf.-Kar. 98k for a Gew. 43 and taking four additional magazines of B-Patronen 'plus some more rounds in his pockets'. Taking cover near wounded men, Allerberger remained unseen:

> This gave him the decisive element of surprise. He let the first two waves of Russians out of their holes, and then suddenly stood up and opened fire at a distance of about 50 to 80m in his well-tried and tested method: always at the enemy who got up last. To be sure of his hits and for maximum effect he fired at the torso, and the explosive bullets ate into the Russians' bellies, tearing apart the abdominal wall and the intestines.
>
> The Soviets were utterly surprised by this fire from the flank and obviously confused. His comrades had in the meantime regained their composure and opened up a well-aimed fire of their own. The Russian attack faltered. After ten shots the magazine of Sepp's self-loading gun was empty. In seconds he had inserted a new one and resumed firing.

> Every shot was a hit. The enemy's fire did not come anywhere near him before twenty Red Army soldiers were on the ground screaming.
>
> He used his third magazine. By this time the screams of the wounded had demoralized the storming party, and they aborted their attack and withdrew to their positions.

However, the Russians soon replied with concentrated small-arms fire. Allerberger seemed to have little chance of escape. So, to protect himself, he:

> pulled the two corpses together for cover and rested his rifle on the thigh of one. Now the hour of the marksman had come.
>
> While his comrades supported him with covering fire from behind, Sepp took the Soviets into his sights from a range of about 100m. With his first two shots he took off the heads of a machine-gun crew, while bullets hit the bodies of his dead comrades and made them twitch eerily. After that it was just like target practice. Every Russian head that appeared over the edge of a trench was shot. Within less than ten minutes another twenty-one Soviets were dead.

Ammunition

When the First World War began, cartridge design had come to the end of a formative period: virtually all armies had small-calibre ammunition – 6.5 to 8mm – loaded with smokeless propellant, and the original round-nose bullets had largely given way to pointed or 'Spitzer' forms which gave better long-range performance. Bullet weights ranged from about 150 to 245 grains while muzzle velocity usually lay between 2,000 and 2,900ft/sec. The German 8mm S-Patrone and the US .30–06 were the most powerful, while the Italian and Japanese 6.5mm rounds offered substantially less power but also more controllable recoil.

When fighting began, and the use of snipers began to increase, the need for specialist equipment grew. Widespread use of iron-plate loophole shields and the introduction of steel helmets in 1915–16 reduced the effectiveness of conventional bullets. Consequently, use was made of high-power sporting rifles, often of large calibre, to defeat the plates, and it was well-known that the 'Dum Dum' effect achieved by filing off or otherwise flattening a bullet nose gave rise to terrible wounds. As Frederick Sleath remarked in *Sniper Jackson*:

> Standing on the fire-step with the sergeant and another sniper, gazing over No-Man's Land. It seemed impossible for them to be seen by the Germans. A bullet sang between officer and sergeant. Their companion slid down into the bottom of the trench with a hole in his head into which a man could have put his fist. 'Tampered Bullet' said the sergeant.

So widespread did the fear of doctored bullets become, even though use was by no means universal, that snipers found with cartridges which had been so treated were often simply executed.

But filed-off bullets could be very effective: in Gallipoli, for example, they proved to be far better at pulverising parapets and loophole-guards. Wilson and Wetherell, writing in 1926 in the *History of the Fifth Light Horse Regiment*, confirmed one use to which 'tampered bullets' could be put:

> To prevent enemy sniping, attention was given to the Turkish loop-holes. These loop-holes on our front were made of green bricks, that is, bricks unburned, but dried in the sun, about 12 in. × 8 in. × 4 in. As soon as these loop-holes were made or repaired within a distance of 400 yards, snipers were put on to them. The heads of the bullets were filed or rubbed off. Such a bullet hitting a brick would smash it to pieces. Five or six shots usually broke up the four bricks forming the loop-hole, which then collapsed and became useless. This procedure was carried on to such an extent that sometimes there was not a serviceable enemy loop-hole facing Chatham's Post within 400 yards of that place.

In addition, reversing bullets so that their base pointed forward sometimes allowed iron plates to be pierced simply by harnessing the so-called 'Munroe effect', which ultimately led to the introduction of specialist anti-armour projectiles.

Armour-piercing ammunition, comprising tungsten-cored bullets, had appeared in German service by the time a document claiming that that *Spitzgeschoss mit Kern* ('S.m.K.') could penetrate 11mm of best-quality armour plate at 100 metres and 4.5mm at 1,200 metres was published on 28 March 1915. Yet the British Army did not follow this important technical lead until the approval of the .303 Mark VII (W) cartridge in July 1918.

Tracer ammunition also appeared during the war, the first successful design being the *Patrone mit Spitzgeschoss mit Langspur (TH)* developed by Goldschmitt of Essen and issued to the German Army in 1917.

Some of the many chroniclers quoted in *Voices of Snipers*. Top row, left to right: Julian Grenfell, Hesketh-Prichard and Herbert McBride. Middle row: Vasiliy Zaytsev, Lyudmila Pavlichenko and Klavdiya Pantleyeva. Bottom row: Yuliya Zhukova, Clifford Shore and George Van Orden.

Michael Ackabee, a First Nations Canadian of the Wabigoon Band, served as a sniper with the Canadian Expeditionary Force in 1916–18. His Ross rifle has been modified to half-stock form, which often improved the shooting of individual guns. (*Courtesy of the Kenora Great War Project*)

Red Army sniper Evdokiya Motina fires an Obr. 91/30 rifle with a PU sight.

General view of a Mosin-Nagant 91/30 sniper rifle with 3.5 × PU sight (*courtesy of Morphy Auctions, www.morphyauctions.com*), and a close-up of the breech showing the one-piece sight mount and the way in which the bolt handle was turned down to pass beneath the body of the sight.
(*John Walter*)

The Red Army 'Excellent Sniper' badge, introduced in 1942 to acknowledge proficiency achieved during training.

Three of the best-known sniper rifles: the M1903 Springfield, the Zf.-Kar. 98k, and the Rifle No. 4 Mk I* (T). *(Courtesy of Morphy Auctions)*

This Obr. 91/30g Mosin-Nagant rifle, with a PEM sight, was used by a Red Army sniper as the stars on the butt testify.

The US M1903A4 Springfield of the Second World War period, shown with (top to bottom) the Unertl, Lyman and Weaver sights. (*Courtesy of Guns & Ammo*)

The German Zf. 4 in its lever-locked mount on a Zf. Gew. 43. (*Courtesy of Morphy Auctions*)

Erected in Manila, Arkansas, this statue commemorates the life of highly successful but largely overlooked US Army sniper Herman Davis, confirming the dates of the principal awards made to him for service in the First World War. (*Courtesy of Thomas Machnitzki*)

This statue commemorating the snipers of the Red Army who gave their lives during the Second World War is part of the 'Monument to the Heroic Defenders of Leningrad'. (*Courtesy of Martin Pegler*)

1

Смерть немецким оккупантам!

БИЛЕТ № 293

УЧАСТНИКА ФРОНТОВОГО СЛЕТА ОТВАЖНЫХ ИСТРЕБИТЕЛЕЙ НЕМЕЦКИХ ОККУПАНТОВ

ТОВ. Николаев Е. А.

Февраль 1942 г.
г. Ленинград.

Политуправление
Ленинградского фронта

This ticket invites Evgeniy Nikolaev to participate in a celebration of snipers held in Leningrad in February 1942. (*Courtesy of Dina Nikolaeva*)

The ill-fated Red Army sniper Roza Shanina, whose diary (see Appendix 2) has ensured that her story has long outlasted her death shortly before the war ended. She carries an Obr. 91/30g Mosin-Nagant fitted with a PU optical sight.

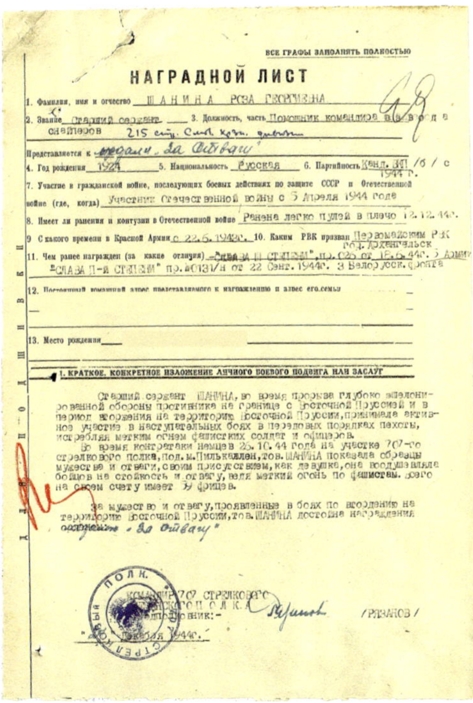

ВСЕ ГРАФЫ ЗАПОЛНЯТЬ ПОЛНОСТЬЮ

НАГРАДНОЙ ЛИСТ

1. Фамилия, имя и отчество ШАНИНА РОЗА ГЕОРГИЕВНА
2. Звание Старший сержант 3. Должность, часть Помощник командира взвода снайперов 215 стр. Смол. Красн. дивизии

Представляется к медали «За отвагу»

4. Год рождения 1924 5. Национальность Русская 6. Партийность Канд. ВКП/б/ с 1944 г.
7. Участие в гражданской войне, последующих боевых действиях по защите СССР и Отечественной войне (где, когда) Участник Отечественной войны с 5 Апреля 1944 года
8. Имеет ли ранения и контузии в Отечественной войне Ранена легко пулей в плечо 12.12.44г.
9. С какого времени в Красной Армии с 22.6.1943г. 10. Каким РВК призван Первомайским РВК гор. Архангельск
11. Чем ранее награжден (за какие отличия) «СЛАВА III СТЕПЕНИ», пр. 026 от 18.6.44г. 5 Армии «СЛАВА II-й СТЕПЕНИ» пр. №0131/н от 22 Сент. 1944г. 3 Белорусск. фронта
12. Постоянный домашний адрес представляемого к награждению и адрес его семьи
13. Место рождения

I. КРАТКОЕ, КОНКРЕТНОЕ ИЗЛОЖЕНИЕ ЛИЧНОГО БОЕВОГО ПОДВИГА ИЛИ ЗАСЛУГ

Старший сержант ШАНИНА, во время прорыва глубоко эшелонированной обороны противника на границе с Восточной Пруссией и в период вторжения на территорию Восточной Пруссии, принимала активное участие в наступательных боях в передовых порядках пехоты, истребляя метким огнем фашистских солдат и офицеров.

Во время контратаки немцев 26.10.44 года на участке 707-го стрелкового полка, под м. Пилькаллен, тов. ШАНИНА показала образцы мужества и отваги, своим присутствием, как девушка, она воодушевляла бойцов на стойкость и отвагу, ведя меткий огонь по фашистам. Всего на своем счету имеет 59 фрицев.

За мужество и отвагу, проявленные в боях по вторжению на территорию Восточной Пруссии, тов. ШАНИНА достойна награждения медалью «За отвагу»

КОМАНДИР 707 СТРЕЛКОВОГО [illegible] ПОЛКА
[illegible] подполковник:- /РЯЗАНОВ/

« » [illegible] 1944г.

Roza Shanina's *Nagradnoy List* gives personal details as well as confirming her military awards.

Roza Shanina wearing the Order of Glory Second Class, awarded on 22 September 1944.

'I received a letter from Arkhangelsk: my friends saw my picture in the newspaper [*Frontovoy yumor*, 7 November 1944], they are writing that they are proud of me.'

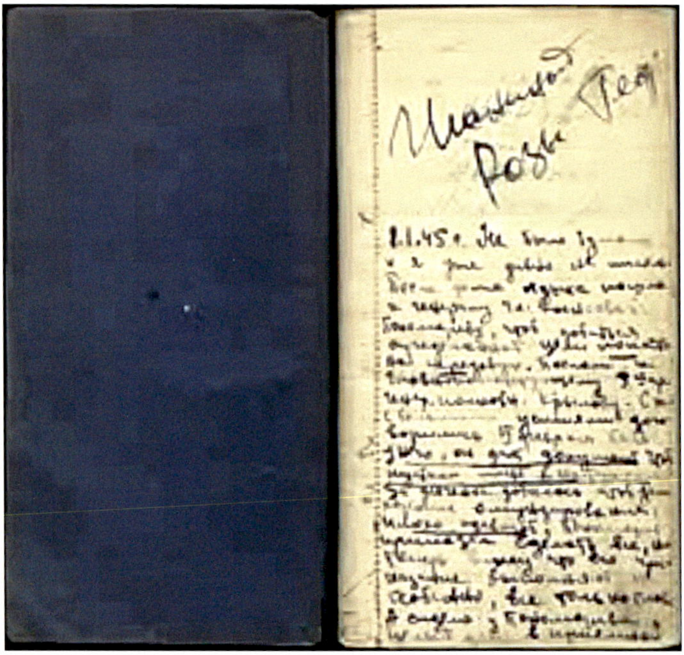

One of the notebooks in which Roza's diary was written, kept by journalist Pyotr Molchanov, editor of *Unichtozhim Vraga*, after her death.

Though useful to observe shot-strikes so that sights could be adjusted (bearing in mind that the trajectory of a tracer bullet diverged from that of a standard projectile), the burning trace could be a liability if it allowed the position from which a marksman had fired to be identified.

Consequently, most snipers used tracer ammunition sparingly – if at all. Vasiliy Zaytsev drew attention to problems with a 'Nazi machine gun':

> I picked up a trench periscope and used it to scan the horizon. These periscopes had excellent optics and were good for observing from fixed positions. I could make out tracer bullets sinking into the cliffs above, with a muffled pounding. I was more than curious about where these bullets were coming from. I had to find their source and take it out of commission – if not now while it was dark, then at the crack of dawn. I should have been able to find the shooter by following the brightly lit tracer bullets, but the machine gun had quickly fallen silent, as if the gunner suspected someone was looking for him.

When shooting recommenced, Vasiliy Zaytsev and his observer Dvoyashin soon saw that wounded Russian soldiers were being targeted. Zaytsev also realised this was the same gunner he had engaged unsuccessfully:

> I was angry with myself for not having eliminated this criminal earlier. I decided there and then to find him and put a halt to his activities.
>
> The fighting started up again at dawn. I changed my position amid the smoke and noise. This enemy gunner had been granted a day's reprieve to keep working his malicious deeds because our snipers in the sector had all been granted twenty-four-hour leave. So now he had the mistaken impression that he could operate with impunity, and he was getting careless. He should never have kept using tracers – they led me straight back to him. I found the scavenger and put him in my cross hairs – his gun fell silent.

The advent of incendiary and explosive bullets prior to 1918 initially had far more influence in aerial warfare than on terrestrial combat. But the Second World War heralded a change, particularly after the German invasion of the USSR in the summer of 1941. A litany of German atrocities, often claimed to have involved women and children, intensified hatred to a point where all vestiges of civilisation were lost. The Red Army responded in kind, particularly when the surge westwards gathered force.

Consequently, snipers on both sides made extensive use of what, owing to horrendous wounds, were described as 'explosive bullets'. Sometimes these were merely standard ogival bullets which had struck at an angle sufficient to 'tumble over' when hitting the target. The German sS-Patrone had this particular tendency, but the Russian heavy-ball round could also do so.

However, the Russians also had B32 armour-piercing-incendiary bullets containing small explosive charges, and the Germans made good use of the *Beobachtungs-Patrone* or 'B-Patrone' which had been designed to act as an observing tool, exploding with a visible flash when striking a hard-enough surface. *Sniper on the Eastern Front* relates the damage that these bullets could inflict on the human body:

> The sergeant touched Sepp's shoulder: 'I think the "fine hat" over there is their leader. If you do him our comrades will get the Ivans'. Sepp knew about the motivation provided by officers fighting in the front line and the demoralizing effect if they fell. In two steps he was alongside the sergeant and had found a good place to rest his rifle, with a clear field of fire towards the occupied trench. To ensure killing his enemy he loaded one of his valuable explosive bullets, which were only rarely found amongst captured Russian ammunition …
>
> Sepp shifted his weapon, taking the small piece of trench into the reticle. He waited for the decisive moment. Then suddenly the fur hat was visible in the telescopic sight. The shot whipped over the distance of 120m in the fraction of a second and struck. Sepp and the sergeant saw the fur hat puff up like a balloon and then burst bloodily like an overripe melon.
>
> Without their leader the Russians became confused and disoriented. Taking advantage of this, the riflemen stormed into the occupied trench and in violent close combat managed to kill all the invaders.

Vasiliy Zaytsev reflected in *Notes of a Russian Sniper* on the death of fellow sniper Sasha Gryazev:

> Sasha had to break cover to heave his first grenade – and he hesitated a second too long before ducking down again. An explosive bullet hit him in the right side of his chest and spun him around like a top. I ran over to him. Sasha knew he was dying. His eyes were already glazing over. He appeared resigned to meeting his fate … We took off his shirt, to try and cover up his gaping wound. It was hopeless. His right arm hung limp and blood was rushing from his torso. We

> had nothing to dress the wound with, so we cut off his *telnyashka* [undershirt, usually striped] to use for a bandage.
>
> Sasha's eyes came alive one last time. He pulled the blood-soaked *telnyashka* from my hands, stuck it on the end of his rifle, staggered to his feet, and brandished his gun in the face of the enemy.
>
> 'Hold me up', he muttered, and we did, so he was facing the enemy's position. 'We're going to win!' he shouted, and then he collapsed into our arms.

However, the perceived use of 'explosive bullets' allowed journalists and propagandists to exploit ridiculous claims. Like many armies, the Germans issued wooden-bullet ammunition, Platzpatronen 33, for use as blanks. The frangible bullet could cause minor tissue damage only to about 25 metres, but Clifford Shore, in *With British Snipers to the Reich*, reported a story 'stranger than fiction':

> That the German snipers were using the 'dastardly' wooden bullet which inflicted terrible wounds! I have before me a newspaper cutting, based on a report which appeared in a medical journal, which reads:– 'Wooden bullets … are invisible in X-ray pictures of the wounds they make. These bullets break on solid structures such as bone, and the scattered fragments are not likely to be located by X-rays. Wooden bullets are said to be effective up to about 100 yards'.

Inappropriate ammunition nevertheless caused problems in individual rifles. For example, Herbert McBride claimed in *A Rifleman Went to War* that many of the difficulties experienced with the Ross rifle during the First World War were due not to:

> some fundamental defect in the rifle itself, but since hearing from members of organizations in the First Division who participated in the earlier battles without noticing any such trouble, I am now inclined to the opinion that it might have been due, in part, at least, to the ammunition.
>
> During the earlier stages of the war the ammunition was all from the old, established factories and arsenals – and it was made strictly according to standard specifications. Later, however, it became necessary to build and equip many other factories in order to keep up the supply, and as these establishments were, necessarily, manned … by new and unskilled workers, and their machines, tools and gauges

> also hastily built, it was no wonder that much of this ammunition would not function properly.
>
> We noticed it particularly in the machine guns. Some brands would not work at all and many others were woefully deficient, causing many stoppages and breakages, often at extremely critical times.

Selecting the wrong ammunition could be problematical, while fluctuations in manufacturing tolerances allowed individual cartridges to feed too loosely while others jammed while entering the chamber.

Vadim Kozhevnikov, contributing to 'Vyssheye strelkovoye obrazovanie' ('A Higher Education in Marksmanship') published in *Pravda* in June 1944, related how:

> Mikhail Petrovich Kondratyuk was observed to be shuffling around some cartridges by the light of the moon, I asked him for a light. 'What are you doing with the cartridges?'
>
> 'Sorting them out', said the soldier, then raised to his ear a fist which was pressed around a cartridge and shook it. Putting the cartridge aside, he explained 'My cartridge clip has a selection: two with heavy bullets, one armour-piercing, one incendiary and five ordinary'.
>
> 'And is this one defective?' 'I suspect so'. And, wiping the selected cartridges with a rag, he declared with due dignity: 'You can rely on our ammunition as a rule. But I'm the finickity sort; if there's a slight dent or the bullet isn't sitting properly, I can't accept it.'

Snipers carried ball, tracer, armour-piercing and 'explosive' cartridges when available. Some loaded only at the last minute, selecting what was considered most appropriate, while others had already filled magazines with mixed rounds. Lyudmila Pavlichenko stated in *Lady Death* that:

> From my belt hung three leather pouches containing cartridges sorted by type: the first with 'Ball L' light bullets, M1908; the second with yellow-tipped 'Ball D' heavy bullets, M1930; and the third with special 'B-32' armour-piercing-incendiary bullets, with a black tip in front of a yellow band.
>
> I took the rifle in my hands and looked through the eyepiece of the telescope sight. The horizontal line covered the figure of the officer descending the steps, approximately down to his waist. I set myself an equation from the course in practical ballistics which we had done at the school and the solution was: distance to the target – 400 metres.

> Placing a cartridge with a light bullet in the chamber, I looked around to select a place to shoot from.

But things did not go quite as she had expected:

> I hit the first target with the third shot and the second one on my fourth attempt, having loaded the rifle with one of the scarce cartridges containing the 'D'-type heavy bullet. I cannot say I was nervous or hesitant. What hesitation can there be after three weeks of desperate retreat under enemy bombs and shells? Still, something got in the way of my concentration. They say this sometimes happens when a sniper first makes the transition from shooting practice at cardboard targets to firing at live enemy.

Boris Godov acknowledged that:

> There were four varieties [of cartridge]: ordinary, armour-piercing, exploding and tracer, and they had different coloured heads – green for tracers, black for armour-piercing, and the rest I don't remember, but they made a fine collection. I was curious as to what cartridges consisted of and, with the aid of a file, I carefully worked away at them – both our own and enemy cartridges. If you file away at a tracer cartridge, white phosphorus seeps out. It glows in the dark and we had fun with it, smearing one another's moustaches while we were asleep. We were idiots; we did not realise that you could poison yourself with it …
>
> The tracer bullets once came in handy for my Mauser. The Germans had dug in at a village and we couldn't dislodge them … Our machine-gunner fired away at the straw roof of the barn, but was unable to set it alight. So, I told him: let me have a go. I fired a German tracer bullet and the barn immediately ignited. Both the barn and the house next to it burnt down.

Difficulties could also arise from differing trajectories: the weight of tracer bullets reduced perceptibly as the trace-pellet was consumed, and the flight paths of the Red Army's light Type L and heavy Type D bullets diverged as range increased. Climate also had an effect. Rem Altshuller recalled that he:

> used the most ordinary cartridges but we were taught to grease them very carefully before going out stalking, and then to wipe them dry.

> In winter we kept them close to our chest because, if you fired a frozen cartridge from a pouch, there would be a significant deviation, especially over long distances.

Bruno Sutkus recalled in *Sniper Ace* that choosing cartridges was of great importance: 'I had special ammunition: the best was captured Yugoslavian and that of German pre-war manufacture, which gave tighter groups.' And Clifford Shore remarked in *With British Snipers to the Reich* that:

> Decently stored and maintained, ammunition very rarely gave any trouble, and with such material misfires were extremely rare. But there appeared to be a tendency to use old and much misused cartridges of all types – Ball, AP, Incendiary and Tracer – for rifle and LMG range practices and although the pantomime-like appearance of tracer always appealed to the average soldier (a child at heart) and I must confess I always got a kick out of it too, it spelled negation to a man's grouping capacity. There were times when I got a decent group firing a clip containing AP, English and American ball and Tracer, but more often than not there was a wide divergence in elevation of the various types.

Accessories

Lyudmila Pavlichenko and her colleagues went into action on one occasion, according to her account in *Lady Death*, wearing:

> Padded jackets and trousers, helmets and forage caps. On top of them we donned autumn camouflage jackets with mustard-coloured hoods and dark-brown patterns on them, and similar trousers, which were quite capacious and tucked into our boots. On our belts we had four leather ammunition pouches, three grenades, a TT pistol in a holster, a Finnish knife in a metal sheath, a small sapper's spade in a cover, a flask of water, also in a cover, and – over our shoulders – a waterproof provisions bag with dry rations for three days (black bread, fatback and a tin of meat stew). Apart from that we took binoculars, torches with batteries and a flare gun.

Rem Altshuller noted that:

> In summer we were not given camouflage gear and fought in ordinary infantry dress. In winter we were issued very warm underwear, a coat, wadded trousers, felt boots, and a camouflage suit with a hood, but not every rifleman was so well equipped. Your rifle was covered with a white wrap and the sight had gauze wound around it. The same applied to the eyepiece because it could flash when you moved and give you away to an enemy marksman. We would take if off once we had got to our positions.

Klavdiya Panteleyeva noted that, in addition to a 'Three Line rifle with an optical sight. Regular one', she had:

> a bayonet in case you had to take part in an attack. There was also an entrenching tool, a mess tin, two grenades, ammo, and a first aid kit. We didn't use binoculars, only the telescope sight. You could see 800 meters very well.

Andrey Ulanov suggested that Nikolay Nadolko was not unduly concerned, because his pack was not too heavy even though he had:

> a rifle, mess tin, flask, a small infantry spade on my belt, underwear, toilet items, a mug and a spoon. True, I often carried my spoon in my foot wrappings, so as not to lose it in combat. It's impossible to carry it in a kirza boot because it jingles around. We had to take a gas-mask with us – we were forbidden to pile it on the cart. We were forced to take it with us. Cartridges and grenades were not supposed to be carried on marches; they had to be issued on the forward line.

Clifford Shore noted that essential items of sniping equipment, though not universally issued, included a No. 4 Mk 1* (T) Lee–Enfield rifle, a Denison smock, a face veil, a scout-regiment telescope, binoculars, a compass, two No. 36 grenades, 50 rounds of .303 ball ammunition, a water bottle, five rounds of tracer and five of armour-piercing ammunition and a ration pack.

And even though many British soldiers regarded the No. 36 as a very real liability when crawling, owing to the chance of unwittingly dislodging the firing pin, Vasiliy Zaytsev confirmed the value that was universally placed on grenades by relating how his commanding officer Kapitan Kotov had once asked for:

> *fyenki* and *dyegtyarevki*, with their shirts off ... He was calling for grenades, *fyenki* were F-1s, while *dyegtyarevki* were RGD grenades

> without their fragmentation sleeves, or 'shirts'. The battalion commander repeated General Chuykov's instructions: 'Grab a dozen grenades and burst in on the enemy together, you and the grenades, that is. I want you both dressed lightly, you without your rucksack, and grenades without their shirts' … The grenade is an excellent weapon. We loaded up with them and took off … The Fritzes really caught it from our 'pocket' artillery that night. We took them completely by surprise, they must have thought they had us on the run, and they weren't anticipating an attack. By morning we had regained most of the positions we had lost the previous day.

Grenades were deemed essential, as they could repel a threat, cover a hasty retreat … or, in extreme circumstances, be a way of avoiding capture. Snipers were particularly vulnerable to terrible repercussions, and there are several cases – including Russians Khusen Andrukhaev, Natalya Kovshova and Mariya Polivanova – where snipers killed themselves by detonating grenades.

Protective equipment was confined largely to the steel-plate shields that had been issued during the First World War to protect trench loopholes and sniper pits. Frederick Crum, quoting his letter written from Ypres on 4 July 1915, reported in *Memoirs of a Rifleman Scout* how he:

> went up to the trenches with ten men, trenches blown away in places, so one has to 'watch it' crossing exposed places. On the way home the Hun started shelling, so I left the trenches and scouted my party round across country. They thought it more exciting than it really was, and thanked me for safe conduct on return. It you watch the habits of the German gunners it assists. We carried home in triumph a heavy iron German loophole plate, which we are having copied (for we are behindhand in these matters).

Protection was also used during the Second World War, as Vasiliy Zaytsev testified. During an attempt to lure a troublesome German sniper into revealing himself, Zaytsev had left a helmet atop the embankment:

> It was a crude tactic, but for the moment it would have to do. If the helmet didn't draw my opponent's fire, I would make a dummy later.
>
> I had scarcely backed away from the helmet when there was a tremendous clang, and the helmet was knocked to the bottom of the trench! By the shape of the dent, I was pretty sure that my opposing sniper had shot from the same position he had used to fire on [his

sniper colleague] Stepan Ivanovich. 'Got you!' I thought, 'If you're so anxious, I'll just have to pacify you before the day's out'.

I went back to surveying the enemy positions with the artillery periscope. At midday I spotted an armoured shield. It was the type of shield our heavy machine gunners had fixed in front of their Maxim guns. The shield was lying upright on the ground, at a distance of six hundred metres, camouflaged behind a bush and some dried grass. Between the branches I could see a black spyhole – the space that had formerly accommodated the Maxim gun's barrel. From time to time the tip of a sniper's rifle muzzle would gleam through the shield's opening.

Sending a bullet through that hole wouldn't accomplish anything. My round would merely ricochet off the sniper's barrel and do nothing more than scare him. I was stuck. I would have to wait until my opponent stood up, or at least raised his head.

Armoured breastplates also appeared, but were unpopular: too heavy, the metalwork could be penetrated not only by the high-velocity bullets fired from sporting rifles impressed in the early stages of hostilities but also by the armour-piercing bullets that subsequently appeared. Yet the idea resurfaced in the Second World War, Clifford Shore noting in *With British Snipers to the Third Reich* uncorroborated German claims that Allied troops assaulting the Normandy beaches on D-Day survived machine-gun fire owing to 'steel linked jackets, or waistcoats'. However, his only personal experience had been:

The three piece shield, chest, pelvic and back, with which certain ack ack units were issued in Normandy in July 1944. This 'equipment' consisted of three 1mm thick, light, manganese steel plates and was designed in 1941 by the Research Council. The breast plate measuring 9" × 8" was to protect the heart and lungs; the 14" × 4" back piece was to cover the lower back and a large portion of the spine, and the third plate, 8" × 6" was to be worn over the upper and central belly. The plates were attached by web-straps and held away from the body by thick rubber padding around the edges.

In 1941 the first impressions of the body armour were favourable and optimistic. But scientific experts declared that it could save only between 1% and 3% casualties, and when field exercises were held to see what effect the wearing of the armour would have on the stamina of the men the optimism changed to pessimism. Rapid and easy

> movements were seriously impaired, and men perspired so profusely that their powers of endurance were affected.

While in Germany, Shore had an opportunity to test a chest shield, firing:

> three shots from a S. & W. .38 [revolver] from five yards distance. Two of the cartridges were of normal British service type – R.L. jacketed, and the other a German cartridge, lead bullet. There was no penetration of the shield, and the dents were by no means severe. The lead bullet was almost completely flattened. At fifty yards bursts from a Sten carbine split the plate, and at 400 yards range every rifle bullet passed cleanly through; shots from a Mauser 7.63 pistol penetrated the shield at a range of 10 yards.

So not everything was as advantageous as it claimed to be!

SHOOT IF YOU CAN!

Using a rifle and an optical sight effectively was clearly vital to any sniper, but decisions often had to be made in the blink of an eye as circumstances changed. Training, if undertaken properly, helped to avoid mistakes which could be fatal. Klavdiya Panteleyeva gave an insight into the process:

> The rifle could shoot two kilometers in a straight line. But you could observe up to 800 meters. At the school we fired at 200, and 300. There was night target practice. Different kinds of shooting, though we didn't shoot at night at the front or in moonlight: we went to our position as soon as it dawned, but returned as soon as it got dark. We stayed not in the trenches, but at the regiment commander's command post.
>
> We only fired one shot from each position, you couldn't do two in case the enemy fired back: we could be killed! Some days we didn't even fire a single shot. In daytime, we stood in the trenches while the soldiers were resting. However, though we weren't supposed to participate in attacks, we did.

Roza Shanina says nothing specific about her firearms training in her diary, but other graduates of the Central Women's Sniping School have been more forthcoming. For example, Yuliya Zhukova, writing in *Girl with a Sniper Rifle*, remembered that:

> The programme included training in tactics, firearms, parade drill, physical development and politics. We were supposed to know by heart the Red Army regulations and the ins and outs of all types of firearms ... We were taught how to set up fox-holes, including reserve holes and decoys; we had to know how to camouflage ourselves and sit in hideouts for lengthy periods, to familiarize ourselves with a new locality and to crawl on our elbows. There were special exercises to

> improve our powers of observation and memory, sharpen our vision and develop firmness of hand …
>
> Usually we would spend the whole day there [on the shooting range], setting off there straight after breakfast and only returning to the barracks in time for supper … We fired at targets from full, waist and chest height – at both moving and stationary targets, open and camouflaged. We fired standing, lying and kneeling, with and without support for the rifle; we fired both on the move and while standing still. All in all, you could not complain that the exercises lacked variety.

And Bruno Sutkus, one of the most successful of the Wehrmacht's snipers, remarked that there was much to remember:

> I used to put out at a hundred metres distance a small label marked with a red spot smaller than my thumb. Then I made sure I could get a five-round group on the red spot. The sniper must be able to estimate distance accurately and gauge the wind speed and direction. Every detail is important … [for example] range always had to be estimated when on the march or in the trenches for there was no automatic-ranging facility available. Target seeking involved inspecting the terrain from top to bottom, left to right and back. I used to keep asking myself, am I looking at a natural feature here, or is it camouflage? Should those leaves be withering like that, the grasses drooping? The Russians were masters of camouflage. Their snipers were mostly from the taiga forests of Siberia and had grown up surrounded by nature.

Sutkus added that shooting could only be mastered if practised endlessly:

> The telescopic sight magnified the target four to six times for the sniper. Targets at 400 metres appeared to be at 100 metres. During aiming one can feel one's heartbeat. As soon as the target appeared at the centre of the crosshairs the trigger was squeezed. The pressure had to be very light so as not to disturb the line of the bullet's flight. It required a great deal of self-confidence to hit the target precisely. It is possible to be accurate up to 800 metres but success is more certain between 100 and 400 metres. There is no point in just shooting for shooting's sake. One has to hit the enemy, or he will hit you.

Few snipers were impervious to error, even after intensive training and combat experience, but identifying your own mistakes and exploiting those made by the enemy was important. Pyotr Belyakov recalled that during a spontaneous debriefing of sniper operations at a Young Communist League meeting:

> Snipers Bazhanov and Khromov criticised their colleague Klimenko for his lack of circumspection. Forgetting to take precautionary measures, he had arrogantly taken up a position in a tree in the open and nearly collected a bullet. Another rifleman, Onishchenko, took poor care of his weapon and failed to grease it on schedule. The reproofs that followed had a sobering effect on the culprits; they subsequently operated more circumspectly and, in time, showed themselves to be excellent snipers in combat.

And Nikolay Ilin reflected:

> How can you get by without making mistakes? But we, like the sappers, must not make mistakes. An enemy sniper once discovered my whereabouts. He fired but didn't hit. I couldn't restrain myself, lost it – fired back with a series of shots based on guess-work. I didn't know exactly where his firing position was. I felt I had made a blunder, but I didn't want to withdraw from such an advantageous position. I had to make the enemy show himself even just for an instant. But how? I resorted to the simplest ploy: I took off my helmet and raised it a little above the trench. The Fritz immediately put a hole through it. I quickly worked out where the flash had appeared, crawled into a reserve trench, and began to wait. The German couldn't hold out. He was dying to know if I was a goner or not. It was a quiet day, with no wind, and I noticed a twig twitching, and then I saw the enemy. I finished him off there and then. In a word, I got away with it, not through patience, but through cunning, and the enemy that time was not too experienced, so I outwitted him.

Ilin then observed that:

> The most criminal thing in our business is to perish from stupidity. There's no end to the number of madcaps and bold spirits who have risked their life for nothing. They strut around under a hail of bullets, and what for? Putting your head in the gun [i.e., being unnecessarily rash] is not on. The point lies elsewhere – to inflict damage on the enemy and remain invulnerable yourself. And how do you do that? My first law has always been, and still is, to cherish the earth. She's

> like a mother to you; she will protect you. Don't begrudge sweat to prepare your position well, and not just one, but several. Camouflage it to blend in with the area. Hard work – that's the secret of success … If you want to be a sniper, learn to determine distances. Rely on your sights, but don't make mistakes, and work it out by eye.

Another problem, especially in the early days, was that few men had sufficient knowledge to make best use of the tools they had been given. Training at OCTU (Officer Cadets Training Unit), Clifford Shore, author of *With British Snipers to the Third Reich*, was one of those unfortunate enough to be issued with old SMLEs and to be told that 'these rifles were no use as regards shooting':

> This surprising statement was amply borne out when we went on to the ranges for the first of our very limited visits. My first three shots at 100 yards were wash-outs, and according to the signal very low and much to the left … For my fifth and last shot, with 600 yards on the sights, I took an aim half-way up the flagpole which was standing at the extreme right-hand side of the butts, and lo and behold, up came a disc to signal a bull.

Finding the Range

Among the most important elements of rifle-shooting was 'zeroing': ensuring that the shot went where the sights had predicted. This had been an age-old problem, beginning as soon as mass production of rifles had begun. Ever-increasing output, and demands made by military authorities, ensured that quantity was the byword. No time could be spared to calibrate each gun individually, and so, at best, random batches were tested to ensure that there was no underlying long-term problem.

Individual marksmen sometimes learned to compensate for their poorly sighted rifles, but the rank-and-file generally blazed away regardless of where bullets were striking. And, in the confusion of a large-scale engagement, it was all but impossible to follow the flight of an individual bullet. Deficiencies in musketry training were still obvious in many armies when the First World War began.

Even after a rifle had been zeroed satisfactorily, periodically checking the settings was obligatory. The slightest knock could move sights by a fraction that meant inches at a hundred yards and feet at a thousand yards: the difference between hit and miss, or life and death.

One of the most important challenges of sniper training was to provide enough variety to prepare marksmen for the battlefield. Shooting at static targets helped to teach use of rifle and sights, but few real targets were static. Disappearing targets were better, particularly if they could move as well as fall. Shooting at replicated enemy trenches taught the value of observation as well as accuracy; 'crawl, observe, consider, shoot' was wise practice.

Work needed to be undertaken at realistic distances, from all angles and under widely differing conditions. From this, of course, little would be learned if the individual could not gauge range satisfactorily. Consequently, it was well worth exploring ways of using ranging or reference marks to prepare for the shot, or even as aim-points for the shot if, for example, the position of a loophole clearly observed with the spotting telescope was not as clearly defined in the less powerful optical sight.

The range of engagement in the trenches of the Western Front and on the heights of Gallipoli during the First World War was often quite short, though differing problems were created in the deserts of Mesopotamia. Regardless of the location, however, a great premium was placed on estimating range accurately: even a minor error could be the difference between piercing a loophole and merely striking its protective shield.

For example, Vasiliy Zaytsev related in *Notes of a Russian Sniper* how he and his partner Kulikov had missed with several shots, after an arduous night-time approach, at a German machine-gun nest only 300 metres away:

> We began fishing for reasons why we could have missed. Maybe the strain was affecting our vision, or our scopes had gone bad, or perhaps our breathing was unsteady, or maybe we had simply forgotten how to press the trigger without jostling the rifle? I looked over at Kulikov. He had dropped his head into his hands, wondering, like me, what had gone wrong …
>
> I was kicking myself. Then I remembered that we had been shooting down at our target. Under those conditions, it's always awkward to measure distance. You can never believe your first estimate – you always have to add on at least 10 per cent of the measured distance, to be safe. And something else – when you've got all kinds of guns firing nearby, the air around heats up and seems to ripple before your eyes. It's a heat mirage, and it makes your target appear closer than it really is. You have to take this into account and either tack on a few metres, or fire a test shot at something close to your target, in order to correctly gauge your distance and zero in.

Fyodor Dyachenko also had problems understanding that:

> bullet deviation presented a particular complexity. I simply had to learn the definition that it just referred to the tendency of a bullet to drift to the right. I could not understand why I needed to know about it. With corrections for wind and temperature everything was plain, so what was the point of going on about 'deviation'? It turned out that adjustments for it also needed to be made when shooting over long distances.

Yet spectacular long-range hits were achieved, though often owing more to chance than calculation. And, it is fair to say, the range of some successful shots grew considerably in the telling!

Most kills obtained during the Second World War were achieved at comparatively short range, just as they had been in 1914–18. Questioned in 1969, Matthäus Hetzenauer (346 kills) suggested limits of engagement to be 'Head up to 400 metres. Breast up to 600 metres. Standing Man up to 700–800 metres'. Sepp Allerberger (257 kills) and Helmut Wirnsberger (64 kills) both considered the limits to be 'Head up to 400 metres. Breast up to 400 metres. Standing up to 600 metres'. Allerberger added that he believed 600 metres to be the limit at which a kill could be guaranteed.

Hetzenauer and Allerberger thought that the ranges they had suggested applied only to the best snipers, while Wirnsberger stated that he, and indeed the majority of snipers, could perform similarly at long range. But he added that a few 'outstanding snipers' could be successful at far greater distances. The extreme practicable range, Matthäus Hetzenauer said, was:

> About 1,000 metres. Standing soldier. Positive hitting not possible, but necessary under the circumstances in order to show enemy that he is not safe even at that distance! Or superior wanted to satisfy himself about capability.

Sepp Allerberger considered 700 metres to be a realistic maximum, while Helmut Wirnsberger said:

> About 600 metres, rarely more. I usually waited until target approached further for better chance of hitting. Also confirmation of successful hit was easier. Used G43 only to about 500 metres because of poor ballistics.

However, Klavdiya Panteleyeva noted that:

> My longest hit was near the Dnepr, a machine gunner and a sniper sitting in a shed, across a field. Probably a kilometre, if not more. A target could be hit up to two kilometres.

Hetzenauer believed his success rate to be 65 per cent hits at ranges up to 400 metres, falling to 30 per cent at 600 metres. Wirnsberger reckoned his percentages to be 80 at 400 metres, falling to 20 at 600 – beyond which, he said, hits could not be confirmed without difficulty. Allerberger said that he did not remember well enough to be certain, though most of his kills were at ranges less than 600 metres.

Into Position

The sniper – indeed, any rifleman – had to learn to fire from many positions, but there was no consensus. Many instructors, such as Hesketh-Prichard, McBride and Shore, were critical of the conventional kneeling position on the grounds that it was not especially stable; and they believed that standing, unless cover could be used beneficially, was rarely an option, as it risked exposing the firer unnecessarily. Herbert McBride said in *A Rifleman Went to War*:

> In any sort of an advance, the type of country and surroundings may undergo a decided change. One moves from the flattened-out country of shell holes and blown down buildings over into comparatively open ground, with growing vegetation and standing buildings – plenty of cover of an entirely different colour and type. What fitted in closely some ten miles back will draw fire if tried in the new surroundings.
>
> Under conditions such as these it will be seldom possible for the rifleman to assume the orthodox firing positions he has learned back home on the rifle range. He must be able to adapt himself to all sorts of restricted positions and still deliver accurate rifle fire. A shell hole or small hollow in the ground will be the usual place from which one fires, and as little as possible of one's anatomy should be exposed to the bullets which will be flying about. Under such conditions, the rifleman invariably rests his rifle over the edge of the shell hole to fire, and one soon learns to take advantage of every possible assistance of this nature … Sometimes it is necessary to kneel or squat, or even stand up, in order to see the target at all. In an advance one often has

> to assume these less secure positions in order to see or fire over top of shelter or cover in the foreground. In thinking it over, I cannot recall that I ever fired any shots from the sitting position – and a good one it is too.

He added that, in firing from trenches, his men's usual method:

> was to stand on the firing step and rest the elbows on top of the parapet, which gave one a very steady position. But the top of the parapet is the aiming point for which everyone in the enemy line generally fires … and one may not last very long in such an exposed situation. So we soon learned to pick out some place where the parapet had been broken down, or we pulled a sandbag or two out of place and piled them up so as to afford some sort of protection from fire coming from the front. Then we would fire out to one of the flanks.

Some championed the traditional back position while others favoured the Hawkins position, widely believed to be suited to slightly rising sites onto which the fore-end could be placed, though often uncomfortable to use as the firer's armpit had to go over the butt.

The use of a sling to steady aim, popular on the ranges in peacetime, was not obviously advantageous in battle. Herbert McBride, writing in *A Rifleman Went to War*, doubted that attaching a sling:

> could be done at all during an advance, or during offensive fighting on a battle-field. It ties down a man too much and hinders his movements and observation, also to be of real assistance it has to be adjusted so closely as to be of use in but one certain position, and you change position very often in a battle.
>
> But the sling is of great help to those in defensive positions, or to those who may get fixed in a good location and stay there. I have seen many old timers in my crowd who had much training on the rifle range, fix their slings for firing and leave them in that position …
>
> The sling is a most valuable aid in accurate aiming and troops should always be instructed in its use.

Special ways were even devised of dragging rifles into position when crawling, while the Germans actively encouraged shooting not only with the rifle fore-end on a rest but also with the left hand cupping the under-edge of the butt. However, using a rest was rarely recommended elsewhere, though trials sometimes showed that sandbags, provided that they had a reasonable degree of resilience, could be very useful.

Though a resilient pad could negate the worst effects, a rigid rest such as a concrete lintel or window-ledge invariably had a disruptive effect on barrel harmonics if the point of contact – usually the underside of the fore-end – bore on the barrel by way of (for example) encircling bands or the nose cap. Only in the case of a free-floating barrel, or one that is precisely bedded for its length, will there be negligible disruption.

The use of trees for cover or as vantage points, which the French and Germans had done widely during the First World War, was decried by most British and US sniper instructors. Many took the view not only that the shot-angle was rarely level, with difficulties of assessing trajectory, but also that the trees gave very poor protection once the position had been detected. Yet there is no doubt that they were widely used. Lyudmila Pavlichenko stated in *Lady Death* that:

> Our regiment had earlier driven the enemy out of the cemetery situated about 200 metres from the southern tip of the village. There were trees growing in the cemetery: five tall maples with spreading crowns and thick ash-grey trunks which had managed somehow to survive the bombing raids and artillery attacks. The difficulties involved in camouflage on the steppe receded here. From the book *Combat in Finland* I knew that in the Karelian forests the Finnish snipers, the so-called 'cuckoos', engaged in targeted shooting at our forces while hidden in the branches of the trees – pines, firs and spruces. Why couldn't I make use of this experience?

Denis Edwards recorded in *The Devil's Own Luck* that, shortly after D-Day, in his area of Normandy:

> Under cover of darkness their snipers had climbed into tall trees and buildings and from daylight onwards began firing their high-powered rifles with deadly accuracy. My first indication was the distant 'crack' as they fired and, almost instantaneously, one of our lads would crash to the ground. They were fantastic marksmen and seldom let off a shot without hitting what they were aiming at. We quickly learned to respect these highly skilled Germans who were well concealed several hundred yards away, yet could pick off a man who only had the top of his head above ground level.

With no ammunition to spare we knew that there was no point in blazing away at every high tree and building in the surrounding area. In any event we were reluctant to disclose our defensive positions and firepower.

And Clifford Shore commented in *With British Snipers to the Reich* that 'a good deal was heard about German snipers in trees', but that he was sure that most riflemen would share his doubts 'as to the efficiency and effectiveness of tree-shooting':

> There certainly were Germans in trees, but I am of the opinion that they were acting as observers rather than snipers. There are comparatively few men who can shoot with sniper-accuracy in the off-hand position, let alone the far more difficult position which arises from almost any stance in a tree.
>
> The Germans used slings for these 'Tarzans' and, of course, such a contrivance captured the attention and fired the imagination of war artists and correspondents, with the result that a most graphic, if not entirely factual, portrayal resulted. The most notable tree merchant I saw – smelled would be a more fitting word since there was little left of him but smell – swung lazily in his sling in a large tree near Cristot in July.
>
> He reminded me of highwaymen's remains swinging from the gibbet in the 'good old days'. A burst of machine gun fire and the hot sun was responsible for this excrescence.

Take Cover!

Concealment was a vital skill, even though battlefield conditions varied tremendously. Herbert McBride, in *A Rifleman Went to War*, related how 'opposite the front of one Canadian battalion':

> there was a particularly clever sniper. For a long time he defied us to discover his lair, all the while taking toll of officers and men. At last, by persistent and systematic observation, he was discovered, but so well was his lair protected that ordinary rifle fire could not silence him. At last the help of the gunners was sought, an eighteen-pounder was detailed to help, and the sniper was driven out, but not killed. I believe his bag up to that time had been two officers and

> seven men. He soon began operations from another well concealed post. This was found by careful and patient observation, and he was again shelled and driven out. Some days later he resumed his sniping, and this time while the Artillery shelled his new post our snipers waited for his appearance and got him. He had probably been worth more to the Germans than a battery of field guns in the same time.

And Clifford Shore drew attention in *With British Snipers to the Reich* to the need to note the difference when under fire of:

> men who had experienced the wiles of the Hun before and the 'fledeglings' [sic] coming up against it for the first time. The majority of men with experience never took chances; they did not require telling to 'dig-in' against the chance of shelling or air attack – it was never too much trouble for them.
>
> The 'novices' at first looked askance at such industry, and ignored friendly advice, but after their baptism of mortar, shell or bomb fire there came a radical change of attitude. Unfortunately for many of them baptism and death were one and the same ceremony …
>
> There could be no two opinions about it – the wise man never hesitated to 'dig-in' and the first job done when one moved to a new locality was to get down to it; fox-holes had priority over grub, and surely no moles were ever more industrious. The first night in Normandy there was not much digging, but more of a 'scraping'. When the Hun bombers came it was a case of diving into the nearest ditch or natural hole, and there was not much thinking about whether such shelters were unhealthy from the point of view of mines or booby traps.

He also commented on the design of sniper-hides encountered as fighting surged across northern France to the borders of the Reich:

> These hides were often interlinked by ditch-ways etc., and in one or two cases it was apparent that one hide had been used as domestic quarters, and the other for the serious job of life. These latter always commanded excellent fields of fire and in the neighbourhood of some of them British and Canadian graves, surmounted by steel helmets which had a single hole through them, were silent, but sufficient, testimony to the efficiency of a real German sharpshooter.

Issue of specialist clothing aided concealment. Red Army sniper Mariya Kataeva recalled that:

> Fighting in winter was particularly hard. The Germans were no fools and donned white trappings like us, so that it was difficult to pick them out against the snow. For the sake of camouflage their officers on the forward line tried to dress like their men. It is only in modern films that they run round in peaked caps and epaulettes. In fact, in their field dress they were barely distinguishable from ordinary German soldiers.

Much has been written about the utility of specific vantage points, whether appropriate or inappropriate, Nevill Armstrong noting in *Fieldcraft, Sniping and Intelligence* that:

> During open warfare snipers must be prepared to take advantage of any position which may help to inflict casualties on the enemy and open the way for the advance of our soldiers, not only in open country when enemy machine guns are holding them up, but also in villages – outlying buildings, rooms in houses, cellars, walls, chimney-pots, and through the roofs of houses which have been partly demolished by shell fire, etc.

Armstrong added that, because 'small villages and small towns are mopped up in more or less quick time':

> Snipers are usually able to move fairly rapidly from one position to another and thereby try to cover the troops who are operating in the streets and among the ground-floor rooms and houses. Snipers should keep a sharp look-out for enemy snipers who may have concealed themselves (very often in cellars) from the first wave of troops passing through villages, and who then emerge and proceed to snipe our men in the back with unpleasant precision and then surrender themselves when they cannot safely kill any more.

Herbert McBride observed in *A Rifleman Went to War*:

> By taking advantage of 'natural' cover, I do not mean to be able to get into some old building and make a nice little nest from which to shoot, although I have spoken a great deal of having done that myself on

> many occasions. When such locations are available, it is fine business – until the enemy gets suspicious and starts to shell the place – but such locations are generally rare and the sniper must be in position to 'carry on' even from the middle of a vacant field. Even when we had our nest in Sniper's Barn, we also had many other places outside – behind a bit of hedge or even just a few sprigs of shrub – once with no concealment other than several unusually long tufts of grass.
>
> An enemy will always devote most of his time to scrutinizing what he considers the likely places for a man to be hiding. It may be a stump, stone, bit of log – anything that would offer concealment. Seldom will he waste time watching what, to him, appears to be an open expanse of ground.

Vladimir Pchelintsev, writing in *Iz boyevogo optya snaypera* ('From the Combat Experience of a Sniper'), related how he had found:

> an abandoned trench under some thick fir branches on a river bank. There was good visibility all round, and an excellent view of the bank opposite and the paths and bushes. I could see everything, but could not be seen myself. I equipped my sniper's post at night. First of all, I made a stout barrier, but left a gap in the front for observation. I deepened the spot where I would have to observe from and lined the walls of the trench with branches. This prevents a trench from caving in and protects the weapons from sand. I made an arm-rest and support from a piece of old board, as well as a couple of forked sticks to lay the rifle on …
>
> Camouflage is very important. I camouflaged the post with particular thoroughness. I planted small fir branches all around and they became covered by snow. I packed the snow in front of the gun-port, so it would not disperse and expose me.

US Army Sergeant Bill Jones reported to Sasser and Roberts (*One Shot – One Kill*) how his only confirmed success depended on what was apparently 'a large dark knot high in the top of a tree':

> I studied it a second and was about to slide on past, thinking it a deformity in the tree, when it moved. I watched it intently, finally concluding that I must have seen a bird, although most of the birds had left. I scoped slowly on down the hedgerow, hoping the sniper would reveal himself.

When nothing else stirred, I returned to the knot for lack of anything more probable. The more I looked at it, the more it resembled a man hugged tight to the tree trunk …

I didn't want the sniper spotting me first if that really was a knot. I studied the length of the hedgerows again before returning to the tree. Several minutes passed … The next time the knot moved. I knew it was not a bird. 'Okay, I have him!' I said triumphantly.

As a light breeze drifted across the field, stirring the blades of grass, Jones:

clicked in a degree of windage and clicked one up on elevation. The workings were smooth. I kept them that way. The good sun provided me a sharp picture through the scope, although the foliage concealed most of what I at first mistook to be a 'knot'.

The sucker had shot at least two of our boys. With that in mind and nothing else, I drew in a deep breath filled with the smell of the rich soil on which I lay, let half of it slowly escape, and then gently stroked the trigger. The rifle recoiled.

When I brought it back down, I saw the tree shaking violently. The 'knot' seemed to be throbbing and pulsating. I quickly bolted in another round and put my cross hairs back on target. Sometimes you had to hit a squirrel two or three times when it was high in the reaches of some sycamore or cottonwood and all you could get the first time was a piece of him.

I squeezed off another round. The German turned loose his perch just like a squirrel does and bounced off the limbs down through the tree until his body hit the ground …

I glanced at the corpse as the company got back in battle formation and straggled across the field and by him. I glanced at him and didn't want to look any harder, but I did. He was just a kid with sandy hair in dirty gray green. His helmet lay beside him. My first bullet must have made the bloody gash across his ribs. The second popped open his chest. His eyes were slitted apart and without luster. His teeth shone dully between his lips. A squirrel's teeth always shone like that when you shot it.

Evgeniy Nikolaev also drew attention to the value of taking cover:

Ivan Karpov and Ivan Dobrik, the best snipers in our division, were … very fond of these buildings – from the top floors there was

> a magnificent view of the German defences. Before daybreak the two friends would clamber into these buildings, make themselves comfortable, camouflage themselves, and observe the enemy.
>
> The Germans meanwhile were preparing for yet another assault on our trenches. Not anticipating any danger from the Klin buildings, which were, if anything, closer to their trenches than to ours, the Nazis were nonchalantly assembling for another attack under the cover of these buildings. The Ivans could hear the Germans loud and clear – playing a mouth organ and bawling their Nazi songs; they had an excellent view of the Fritzes' self-satisfied mugs through the telescopic sights of their rifles. Orders were being given by an officer with a monocle in one eye and a riding crop in hands which were clad in black gloves. The Nazi mob was only about three hundred metres away.

Karpov was the first to fire, felling the officer with the monocle. The two snipers then began to eliminate the others:

> Clearly set up, as they were, for the gunsights of our exterminators. Having worked out what was happening and finally guessed the source of the lead bees with mortal stings, the others began scurrying round the gully, looking for suitable cover. The soldiers were no longer up to singing; without leadership, they fled in panic in various directions. But the accurate firing of the two Ivans caught up with them, wherever they were …
>
> That day … Dobrik wiped out eighteen of the Nazi butchers and Karpov, fourteen. But the main thing was that on this occasion they frustrated an operation on which the Fritzes had placed great hopes.

However, believing sniping from buildings to be dangerous, 'not so much the actual firing … but in approach work', Nevill Armstrong reminded his readers that:

> One sniper could cover any movement up a long open street without any turnings in it, and that by two snipers working together, one on either side of the street, severe casualties could be inflicted. It is therefore far better to select an approach by a back door instead of the front – much more cover may usually be found on the premises behind houses than in front. Only the most expert snipers should be detailed for this kind of work.

There were many other ways of taking cover, often successfully created by individual snipers on the spur of the moment with the most unlikely props but sometimes owing more to assumptions made by onlookers. For example, Clifford Shore suggested that in Normandy after D-Day:

> We heard, and read, much about snipers in church steeples, and one evening I was solemnly told by three men cowering in a ditch that a sniper was firing at them from the church spire which could be seen in the distance towering above some high trees.
>
> This spire was about 1,000 yards away, but the range according to the 'sectional average' of the three men was 'just about 500 yards'!
>
> I am told that the Hun's native ingenuity once gave rise to a 'sniper in a church steeple' true story; to a long length of chemically treated rope suspended inside a church spire the Germans tied a number of 'crackers' at irregular intervals, and then set a light to the rope which acted as a slow burning fuse. The effect of the crackers going-off at irregular intervals suggested sporadic rifle fire – to the credulous, 'sniper fire'!

In October 1942, Evgeniy Nikolaev had set up a firing position:

> under a tramcar which had been wrecked by the Nazis – the same one shown in the film 'Blockade' based on Andrei Chakovsky's novel of the same name. The tram had been stopped by the Nazis and its passengers killed. It stood on no-man's land, but closer to our trenches. At night my friends helped me to prepare a firing position and from first light I began to direct my lethal fire against the Nazis.
>
> It was a successful day for me; I wiped out eleven Fritzes. As we subsequently found out from prisoners, I killed a general, two colonels, who had come to inspect the forward line and, along with them, four officers and a motor-cyclist who were accompanying the higher ranked dignitaries. The rest were ordinary soldiers. Their motor-bike and side-car were consumed by fire.

But, as he subsequently admitted, Nikolaev had been carried away by his success, and had not paused to consider that he would be subjected to counter-fire not only from riflemen but also artillery:

> After all, the isolated tramcar on a completely bare patch of ground was an excellent reference point! I only considered this when I heard

> the howl of a shell flying in from afar. I was able to hear and see the explosions 100–200 metres from the tram. I paid dearly for my thoughtlessness; as my friends told me several days later, the eleventh shell exploded almost right next to the tramcar. I was covered with huge clumps of earth and my rifle, which had been lying on the parapet, was damaged beyond repair, while I was dug out by my friends barely alive.

But there were countless times when taking cover meant the difference between life and death. Klavdiya Panteleyeva noted that:

> Germans attacked our trench and took some girls prisoner, and killed them there. They killed Klava Monakhova. Only one soldier survived, there was an abandoned dug-out, simply a hole in the soil covered with a ground-sheet with snow on top … so he spent the day there …
>
> Apart from excellent marksmanship, you had to hide yourself very well. They really pushed us at the school because of that. Sometimes you would sit down, but your entire body could be seen. You had to camouflage yourself so you couldn't be seen. Fit your surroundings. When we arrived during the snow, they gave us special coveralls. We had green camouflage in summer, but there were no blotched versions. They gave us green pants, green tunics. We were always wearing pants, not skirts. Winter pants in the winter, summer pants in the summer.

Bruno Sutkus noted that, in 1944, because his battalion had lost a number of men to Russian snipers, he had been ordered to investigate:

> My first act … was to survey the terrain very precisely and decide where I would have chosen to be were I in the boots of our Soviet opponent.
>
> A sniper does not fire directly ahead but from an angle of about forty degrees in order to avoid rapid detection. One must have good cover and be well camouflaged in a position from where one has the best possible view of the enemy trenches. The range must be estimated as exactly as possible to ensure that the bullet does not fly too high or too low. Some snipers used tracer bullets. This was crazy, for it gave the enemy warning and enabled the source of fire to be pinpointed quicker.
>
> The sniper had to avoid locating himself near an imposing feature – such as a large tree – which the enemy could recognise quickly to focus his search. The camouflage had to be chosen carefully to accord

with nature. One must always think that the enemy sniper is one's equal, and always practise care and self-control, and remain confident.

Nothing on one's clothing should present the enemy sniper with something to aim at. The most minor mistake would be enough to provide the enemy with a target. Upon sighting a target, one has to determine that it is a target and not a lure, then aim lightning fast – quicker than the opponent – to keep living.

Hide and Seek

One of the most important lessons a sniper could learn was to hide himself effectively. Eric Penberthy wrote in *The English Review* (1920) that at British Army sniping schools, men who proved to be good shots:

> were shown how to construct and conceal snipers' posts – some for shooting from and others purely for observation; and because when they were in the line this work had to be done at night, much of the training was carried out in darkness and perfect silence …
>
> The use of 'camouflage' was also practically illustrated. Demonstrations were given in the use of such devices as snipers' suits painted to merge into grounds of different colours; dummy heads for exposing above the trenches and drawing Hun sniper's fire in order to detect his position by flash or smoke of his rifle; white suits for use by night patrols when snow lay on the ground, and many other clever ideas for protecting the sniper or observer when he had to expose himself to enemy view. The Boche, of course, used many similar devices, but I think we beat him at the game.

Nevill Armstrong noted in the 1941 edition of *Fieldcraft, Sniping and Intelligence* that snipers must not only have a good eye for ground and be able to read it quickly, but also to know how to stalk successfully should such an opportunity arise:

> This is a very important part of a sniper's training. They cannot have too much of it, whether they are operating in trench warfare or open warfare.
>
> Great stress should be laid on watching the actions of animals and birds at all times, and noting how frequently the position of human beings is given away by the actions of animals, etc. Also

> protective colour is not sufficiently understood by snipers, scouts and observers.
>
> In position warfare we are very much like animals. We do the same thing more or less in the same way day after day. Therefore a knowledge of the habits of animals is quite an asset to begin with.

Martin Pegler recalled that during his own British Army sniper training, the advisability of looking at trees, bushes and grass every ten minutes was emphasised, in the belief that green and brown colours relaxed the eye muscles.

An idea of the way in which environment could be obscured can be gained from Hesketh-Prichard's remarks on loopholes in trench parapets:

> The efforts to camouflage our loopholes were extraordinarily primitive – indeed, concealment was nearly impossible in the form of parapet then in use. Many of our units took an actual pride in having an absolutely flat and even parapet, which gave the Germans every opportunity of spotting the smallest movement. The parapets were made of sandbags beaten down with spades, and it is not too much to say that along many of them a mouse could not move without being observed by the most moderate-sighted German sniper … At a later date a trial was instituted, and proved that in spotting and shooting at a dummy head exposed for two to four seconds over a flat parapet, the number of hits was three to one, as compared with the same exposure when made over an imitation German parapet.

He also observed that:

> The German trenches were deeper, with much more wire in front, and … looked like the course of a giant mole which had flung up uneven heaps of earth. Here and there, a huge piece of corrugated iron would be flung upon the parapet and pinned there with a stake … Here and there lay great piles of sandbags, black, red, green, striped, blue, dazzling our eyes.

And Wilson and Wetherell, writing in the *History of the Fifth Light Horse Regiment*, recorded that in Gallipoli the unit had:

> occupied the extreme right flank on Harris Ridge, and special attention was given to the matter of sniping and breaking in of the enemy's loop-

> holes. Special men were set off as snipers, and taken off other duties to enable them to give exclusive attention to their special job.
>
> Special attention was also given to the Turkish periscopes. The moment they appeared they were fired on, and smashed. This smashing of loop-holes and breaking of periscopes resulted in our people being almost free from enemy snipers.

The Turks were not averse to improving their positions, which meant additional attention was required to neutralise them. Trooper Allan Campbell of 2nd Light Horse, who spent three months in a six-man team of snipers operating from Pope's Hill, subsequently reflected that:

> Our possie would be from 200 to 300 yards from the Turkish trenches, which had a good earthen background that made sighting their heads difficult for us, particularly with much rubbish and debris strewn around to add to our difficulties of sighting them.
>
> For this reason we worked in permanent pairs, one man spotting with a telescope who directed his mate's fire. It was marvellous how quickly we understood each other's directions and how very accurately we could control the fire of our mate. This was one good reason why our pairs were permanent so that understanding was preserved. Our routine was – each pair on duty for two hours with four hours off duty, day and night. Our sniping couples would alternate the jobs of spotting and shooting.
>
> After a while the Turks would build, at night, small tunnels across their parapet, with sandbags to protect their heads, while shooting down our main track of communication to the Beach. This meant more work for us. With daylight we would have to destroy those tunnels.
>
> Sometimes, after a good shot, we would see a hand scooping the sand from a tunnel and have a shot at it. Usually he would just wave us a miss.

The Turks subsequently abandoned the widespread use of sandbags on the grounds that they were too easily destroyed, and built their tunnels from concrete blocks which:

> were much more difficult for us to destroy. We would place several shots as close together as possible which would have the desired effect. Once the main block had been broken then the smaller pieces were quite easy to break.

> For destroying the tunnels, we used to clip the nose off our bullets – but that tends to upset the accurate flight of the bullet. Being against the 'rules of war' we never carried such bullets about, nor had them on hand, because if you happened to get captured with them on you or in your possession you were likely to be shot on the spot …
>
> We used the Mark 6 rifle ammunition with round nose bullets. We did not care for the Mark 7 with the sharp, pointed bullets that the English soldiers had.

Trooper Ion Idriess of the 5th Light Horse Regiment, described sniping at Gallipoli in diary entries made in 1915:

> The Turks have a small trench … [above Monash Valley] which commands our road. Their trench is filled with snipers, splendid shots, who have killed hundreds of our men on this road [27 May].
>
> [Turkish] Snipers got fifteen … It's wretched to think of so many men getting killed and maimed when they are not in the firing line … [28 May].
>
> This trench sniping is intriguing. I am a crack shot, and was put on sniping. Not seeing any Turks I blazed away persistently at one of their loopholes. The invitation was presently answered. The sandbags along the trench are all spattered with bloodstains … [29 May].
>
> I've just been indulging in a duel with a Turk, shot for shot. I'd fire, and the dust would fly up against his loophole … and his bullet would plonk into the sandbag above my loophole. Then my turn … It was thrilling. I waited for each of my turns with every sense keyed to concert pitch, thrilled through and through. No doubt Johnny the Turk felt the same [29 August].

From the earliest days of their craft, snipers have always made use of camouflage and cover – ranging from leaves and twigs woven into scrim stretched over a helmet, and rags tied round a rifle to break its silhouette, to purpose-designed clothing such as the *Tarnjacke* and that descendant of a 'ghillie suit' known as the Denison smock.

Paleness of face and hands can be hidden with a veil or dulled with ash, charcoal, mud, and a man's outline can be disrupted by loose garments or padding. Properly attired, as many battlefield reports testify, the sniper can lie, undetected and undetectable, no more than a few yards from his enemy.

Commanding 3rd Battalion AIF, Lieutenant Colonel Ernest Brown reported that one sniper was 'a Turk in civilian clothes', while others were 'wearing Australian uniforms' or 'dressed in green clothes, with face and hands stained the same colour'. Private Arthur Greenwood of 8th Battalion captured a Turkish sniper in Gallipoli in February 1916. He subsequently wrote how the man:

> was concealed in the scrub decorated as you see him. You could not see him in daytime he being exactly like a bush and not being able to see the flare of his bayonet only by night.
>
> He was getting a lot of our men all the time. In fact I don't know how he missed me it was not for want of trying …
>
> Well how we got him. We crept out, snakelike, worked right round him, not knowing any minute we might be falling into their hands, but we were lucky I suppose, however we got to [the] mark alright, after a lot of manoeuvring.
>
> We got up within close range, concealed ourselves, waited for the moon, as we did not intend killing him so long as he stayed there which he did. You can bet your boots he would not have done had he known he was so near to his doom. As soon as the moon gave us enough light we began operations. I was to disable him while the other chaps covered him in case he attempted to get away.
>
> As he was taking steady aim at someone in our trenches I was doing likewise to him. I was to shoot him through the hand so as to disable him from doing any damage with his rifle then I was to creep up and capture him while the other chaps kept him covered.
>
> We got so far so good but we had the biggest job now to come. We had to get him back to our lines without being detected by the Turks but we got back with our bundle as you can see.

Clifford Shore described in *With British Snipers to the Reich* how:

> Whilst some instruction was given, and camouflaged denim suits issued, the camouflage of weapon and person was left to the individual; by this it was held that a man would be instilled with a self confidence which would not be gained if a certain type and design of dress was forced upon him. (In Italy it was of course much warmer than in N.W. Europe where snipers were happy to wear their Denison smocks in both winter and summer.)
>
> Occasionally the snipers went out on a tour looking like shabby old poachers, carrying only their sniper rifles and no equipment.

Harry Furness described how his Denison smock had:

> Large pockets in which we kept our prismatic compass, notebook from which we later wrote (always in capital lettering so that it could be easily read by the intelligence office) the patrol details. We had a survey map of the area in which we were operating. During the winter snow and frost I also had a white snow suit to wear, and I wrapped my rifle and telescopic sight in white bandages. Later in the campaign I was issued with a complete camouflage suit, a smock and trousers of a very unusual disruptive pattern. It had very mellow colouring and was in addition windproof so I found it very useful indeed as snipers get very cold. Frequently over my boots and gaiters I found it helped if I slipped an unused sandbag and tied it with string, it disguised the outline of my boots and was a help in preventing me slipping if I hurried over smooth pavements with my steel studded boots, it also muffled the sound. Camouflage cream was available at some Sniper Schools but not often in the frontline so we just wet some mud and smeared it on our face and hands.

And Lyudmila Pavlichenko prepared herself for a mission against a German machine-gun emplacement hidden among trees in a graveyard by spending an entire evening embroidering her:

> new camouflage jacket with its green hood and brown pattern. The sergeant major had given me some shreds of camouflage netting and somebody's old tunic, which I cut up into ribbons and short strips. Shaggy ribbon came in handy for covering the rifle barrel.
>
> I distributed the rest of it, along with maple leaves, twigs and clumps of grass, so that it [the jacket] lost its former clear outline and began to resemble the garb of a wood sprite or a marsh devil.

New Tricks

Among the aids available to the First World War sniper was the 'Hesketh Head' – sometimes sporting a dummy cigarette attached to a long rubber tube through which smoke could be blown. The head could slide vertically on a plank or similar support, allowing long-enough exposure to be shot by a German sniper.

The bullet passed harmlessly through the head, which was then lowered by the precise height of a trench periscope. Looking through the bullet holes from the rear, into the ocular lens of the periscope, aligned the objective lens directly with the path of the bullet. The operator could see exactly where the shot had come from, promoting effective counter-sniping. Trials showed that of 71 'heads shot', 67 revealed the German sniper's position. Frederick Crum recalled an incident in 1916:

> When a Brigadier of the 121st Infantry Brigade, occupying trenches at Ploegsteert, suffered so many casualties from Hun snipers that he came up to the Second Army Sniping School and asked us if we could do anything for him, so we gave special training to two of his officers and sixteen men and sent them off. Some two weeks later we received a letter from this Brigadier, stating that the sniping had been checked and the tables turned, and that one of the most dangerous and persistent snipers had been spotted by using a camouflaged head, two of which we gave the sniping officer and trained him in their use.
>
> The head was carefully exposed and the Hun immediately bit, but missed with his first shot; the second shot, however, hit the dummy head in the eye and came out behind. By looking through the holes with a periscope the Boche was spotted in a tree and killed.

But not everything worked as planned. Crum also noted in *Memoirs of a Rifleman Scout* how on his sniping and scouting course:

> Observation and other practices were carried out with men dressed up in German uniforms occupying a reproduction of the German trenches. Targets were located by observers working in co-operation with snipers and trench mortar officers, who at once opened fire.
>
> Then came the acting, which caused great interest and amusement, combined with instruction. Some of the men had great talent that way, and their topical jokes went down – but when it came to the loophole scene ...!
>
> Two young soldiers were mishandling a periscope, doing everything wrong, when one of them saw a Hun showing his head in the German trench opposite. Getting excited he entered the loophole, sacred to snipers, loosed off 10 rounds rapid, and thus, after having clearly indicated the position of the loophole to Fritz, he retired from it leaving the shutter and screen open, and assuring his companion that he had exterminated the lot.
>
> Presently the sniper came along the trench and inquired what all the firing was. Being suspicious, after a lecture on 'giving away' loopholes, and the dangerous work snipers had in building and concealing them,

> he had just cautiously entered the loophole when a shot rang out from the German trench and a yell of pain came from inside. Out came the furious sniper.

Typical of the ways in which individual snipers were required to improvise was one, as detailed by the Nikolaevs, in which Pyotr Goncharov:

> went off to his firing position, taking sniper Konovalov as his partner. For three days they crawled around the steppe in the thawing March snow, seeking out the enemy. But, as if sensing that danger was close, the Fritzes lay in their snipers' nests, not giving themselves away with a single shot …
>
> Goncharov could not go back to the battalion empty-handed. He thought up dozens of ploys: planted dummies, fired at soldiers who appeared on the other side and wiped out seven Nazis in three days, but the German snipers remained silent. Then Goncharov decided to embark on the ultimate and most frightening stratagem possible in a sniper duel. He crawled the one hundred and fifty or two hundred metres towards the line of German trenches and, burrowing into the snow-drifts, began to wait. He knew that the snipers would not be able to sit in their lairs for ever, that the frost would eventually drive them out, and they would start moving back to their own rear. That was the moment he must not miss.

On this occasion, the Russian sniper's assumptions proved to be right:

> On the fourth day the German snipers began to leave their hide-outs. In the frosty air by the Donets Goncharov fired the five accurate shots which he had been waiting four days to discharge. Only a single shot was returned by a Nazi sniper, who had discovered Goncharov. The bullet passed within half a centimetre of his temple and left a hole in his cap. However, Goncharov's bullet in response hit the Nazi half a finger's width above his ginger eyebrow. Only one of the six managed to evade Pyotr Goncharov's rifle.

Graduation

Once sniper-training courses had been properly established, some method of identifying successful trainees became necessary. This ranged from simple certificates in the earliest days, such as those conferred by Hesketh-Prichard, to badges, mementoes and even weaponry.

Bruno Sutkus remarked that at the end of the course he had undertaken in the sniper school in Vilnius:

> All those who qualified received the most modern sniper's rifle with telescopic sight, binoculars and a camouflage jacket (*tarnjacke*). I was given a certificate confirming that I had passed out from sniper school and was warned never to allow another person to handle my rifle

In October 1942, Nina Lobkovskaya had been sent to the newly formed women's sniper school in the village of Veshnyaki, near Moscow. She recalled how the winter of 1942 had been exceptionally severe, yet every day the studies lasted for 10–12 hours. At last, nine months later, the field day:

> We had been dreaming so long of arrived at last. In the glittering sun that was reflected in windows, with duffle bags, sniper rifles and rolled overcoats behind our shoulders, we marched through the streets of Moscow to the Riga Railway Station, singing 'A Farewell to Our Beloved City' … Our group of fifty girl snipers was assigned to the 3rd crack army on the Kalinin front.

At the special graduation ceremony:

> Over the months of training the girls grew close to their school, took a liking to their demanding officers and instructors, and became particularly attached to the head of the political section, Major Nikiforova …
>
> On completion of the course Major Nikiforova accompanied the girls to their military detachments and handed each of them a sniper's record book, in which they would record their personal tally against the enemy.

In addition, high-ranking representatives of Komsomol gave elegant Finnish knives to the graduates as a memento. The best shots received presentation sniper rifles with the names of the new owners engraved on metal plates screwed to the butts.

A Finnish soldier pictured during the Winter War of 1940 carrying a Soviet Obr. 91/30g Mosin-Nagant fitted with a PE scope. Large numbers of these rifles were captured and turned against the Red Army.

Red Army sniper Abukhazhi Idrisov (349 kills) poses with his SNT-40 (Tokarev).

A Canadian sniper, often named as Jack Bailey (but not confirmed), takes aim. The presence of a long-range sight dial plate on the left side of the fore-end shows that he has a No. 3 Mk I* (T) rifle fitted with a Warner & Swasey prismatic sight.

A Canadian soldier takes aim with a Mauser sporting rifle 'somewhere in Germany' early in 1945.

This photograph shows Red Army sniper Evgeniy Nikolaev in position, well camouflaged though the optical sight on his Obr. 91/30g Mosin-Nagant can still be discerned. (*Courtesy of Dina Nikolaeva*)

The ultimate rehabilitation – Evgeniy Nikolaev, author of *Red Army Sniper* and one of the highest-scoring Red Army marksmen (324 kills), pictured in the 1990s with his awards and decorations at a time when the performance of the snipers was finally being widely acknowledged. (*Courtesy of Dina Nikolaeva*)

SNIPERS IN COMBAT

The nature of the world wars meant that snipers served in areas which could be hundreds – indeed thousands – of miles from their homeland. But first they had to get there.

German marksmen were parachuted into Crete in 1941, while those involved in amphibious landings included island-hopping US Army and Marine snipers in the Pacific, Allied men landing at Salerno and Anzio in 1943, and others who waded ashore on D-Day in June 1944.

Getting from North America to Europe, North Africa or the Pacific involved sea travel, which often came as a shock to men who, in some cases, had never even seen the sea. Clifford Shore reflected in *With British Snipers to the Reich* how, on D-Day, the landing craft taking him to the beach-head had been attacked by Junkers Ju-88s and:

> was leaking like a sieve … when we took up 'Abandon ship stations' we found the floats useless and everything very badly knocked about. I had the bad luck to lose one or two of my best men, killed and wounded. I can never forget that night; the sense of oppression; the choking in the throat from the smoke generators which had been lighted on deck to 'blanket' the craft from the Hun planes; the demon wailing of hundreds of ships' sirens; the headlong rush for the sparse cover; the bowel-relaxing suspense of waiting within a thin steel chamber for the inevitable attack; the sudden roar of diving planes; the scream of falling bombs; the convulsive lurch of the ship as the bombs tore the water and the sides of the craft; the tearing of steel and the moans of the wounded with the pearls of shock-sweat on their foreheads! It was the longest night of my life.

Snipers such as Irishman Paddy Devlin were among the paratroops and glider-borne Allied forces involved in the ill-fated attack on Arnhem. However, most snipers involved in the war reached their destinations by train, truck or simply by covering long distances on foot. Yuliya Zhukova noted in *Girl with a Sniper Rifle* that:

> Ahead lay the front. The train was carrying us westward. The 'us' represented a whole trainload of girls who were mostly scarcely more than twenty or twenty-two years old. Our train consisted of a long line of heated goods waggons, which had been used before the war to carry cargo and livestock. Now they were carrying soldiers. The waggons had bunks like the barracks, but no pillows or other sleeping gear. I don't remember precisely, but I think we slept on bare boards, spreading one half of our overcoats beneath us and covering ourselves with the other.
>
> In the middle of the carriage stood a cast-iron pot-belly stove. It provided a little warmth and we used it for cooking and making tea. For the whole journey we relied on concentrates, from which we boiled a sort of cross between a soup and a porridge. We had rusks in place of bread and salted herring instead of meat. When we looked wistfully into our mess tins containing the same pottage day after day, or chewed the black rusks, or sucked the herring bones, we invariably recalled the lavish catering at the school and sighed in sorrow.
>
> I do not remember what we felt at the time, how various people behaved, whether the girls were tearful or distressed. But I do remember other things: the way we sang all our school songs one after another and sat for long periods by the open waggon door.

Finally, Zhukova's unit reached the front line and settled in a:

> long, deep, trench, almost the depth of a man's height, equipped everywhere with firing positions and observation points. From the trench, communication passages just as deep led off to the rear. Between our positions and those of the enemy lay no-man's land. But the Germans were almost next door – you could get a good view of them through the telescopic sights of your rifle. Sometimes a helmet would rise above the parapet on one side or the other. In the evenings one could hear the enemy singing and playing mouth organs. Those on the other side had a life of their own too.

There was a certain thrill, she said, about reaching the firing position:

> Excitement – but also insecurity, expectation of something unusual and, of course, fear. Back then during the early days I was afraid, in my

own case, not so much of the Germans as of self-doubt; what if I don't know what to do, can't cope, make a blunder, become a laughing stock? Real fear of the enemy came later, when we were at close quarters with them and had begun to go out on assignments and take part in the fighting. But gradually I managed to overcome these doubts and fears. It would be more correct to say that we learned somehow to dismiss fear, to stop it dominating our feelings and will and depriving us of the ability to think clearly and act responsively.

Klavdiya Panteleyeva was sent to serve with the 1156th Regiment of the 344th Rifle Division, 33rd Army, in 'cattle cars with stoves':

> They couldn't get us all the way to the front, unloaded us. There was such a snowstorm, they gave us a truck to bring us closer to the front, to a reserve regiment. A truck! We carried it all the way on our backs, there was so much snow. So we got there. I don't remember how long it took, a day, two, three … It was a long time ago. They gave us camouflage coveralls. We wrapped bandages around our rifles. Early in the morning they fed us and gave us sandwiches to take with us: bread and American sausage. That was the entire dinner! And we went to the trenches. Everything was packed with snow.

Panteleyeva then described her first experience of the realities of war:

> There was a lake in one place. Germans went there to wash, even in their underwear. So Zina Gavrilova fired, shot one. And Germans stopped going there to wash. We were already on the defensive, it was summer, May or June, and not everyone stood watch at her embrasure, because there was no movement among the Germans, or our movement either.
>
> It was already in the evening, but Germans just couldn't be dislodged from their trenches. They told us girls to go there as well, also drivers and locals. Why? We reached the trench, but couldn't do anything, it was already getting dark. There remained very few people, just the girls and drivers. They told us to take all the wounded we could carry and return to our trenches. We couldn't take them all because there weren't enough of us. Germans finished off those that remained, they screamed so much! – they were bayoneting them to death.
>
> They left us in our trench throughout the night. All minefields were cleared in front of us. I stood the watch there, could see nothing.

> Others were also standing watch, but I couldn't see them. I got so tired by 4 AM, couldn't take it anymore.

Nina Lobkovskaya described her initial experiences:

> When we arrived, we spent the days in constant observation of the enemy's frontline. When we came back for a night and went to bed, our memory held the picture of the terrain in every detail, with every leaf, every blade of grass standing out. And, as we arrived back to our observation positions the following day, we noticed the slightest of changes. This observation technique we had acquired at the sniper school proved very useful …

Where 'creature comforts' were concerned – food, clothing, soap, paper, books, sleep – snipers, like many of their colleagues, had to make do with what could be found: which was often practically nothing, given the scale of destruction in most war zones. Klavdiya Panteleyeva noted that as:

> there was nowhere to sleep, we slept on the ground. And the bed – Marusiya and I put our padded jackets under us. Everyone was freezing. An accordion player once said: 'Why don't you all dance, to get warm.'
>
> Once we found some house, it was completely empty. Everyone immediately lay down, but there was no room for me. There was one trough, a small one, used to chop cabbage. My choices were outside, on the second floor, or that trough. I lay down in the trough: I was small (157 centimeters tall) and skinny. It was uncomfortable. I would stretch my leg, someone would push me back, stretch my arm, someone would push me back again.
>
> Sleep wouldn't come, but I wanted to sleep anyway. In the morning someone started leaving, so I dived for that spot. Slept for a little while, and then we had to get up.

While Antonina Kotlyarova remembered that it was difficult to wash 'yourself and your clothes at the front':

> Once we entered some village, there was a bathhouse there. We went there and barely got out afterward – carbon monoxide poisoning, we didn't know how to operate it. Basically, it was difficult. But I never had any lice. Soldiers had them. They had close haircuts. And at school they also gave us haircuts like the boys.

Supplying food was also often difficult under combat conditions. Yet Anatoly Navara regarded his unit as:

> well fed, but with occasional lapses. There were periods of several weeks when we ate nothing but swede. And once an officer discovered several wagon-loads of buckwheat under an embankment, and for a certain period that is all we ate: buckwheat porridge for breakfast, buckwheat soup for dinner, and buckwheat porridge for supper. After a week of this diet, we were dreaming of swede. But despite the disruptions, we could always rely on getting tea with sugar and a small lump of butter in the morning.
>
> Supper was the only meal we ate in the trenches; breakfast and dinner we took with us as dry rations. They included American corned beef, I remember this, and big lumps of sugar which we smashed with hammers. There were also American canned vegetables. These were only issued on the forward line. When we went to the front, a field kitchen cooked borshch and porridge, but only fed us twice a day instead of three times: at dawn and in the evening.

Into Action

Once at the front, after acclimatising as best they could, snipers were sent into action as quickly as possible. But few of them were prepared for the diversity of their tasks. A memorial pamphlet, privately printed by Viola Meynell after he was killed, details Julian Grenfell recalling:

> I longed to be able to say that I liked it, after all one has heard of being under fire for the first time. But it was beastly. I pretended to myself for a bit that I liked it, but it was no good, it only made one careless and unwatchful and self-absorbed; but when one acknowledged to oneself that it was beastly, one became all right again, and cool. After the firing had slackened, we advanced again a bit into the next group of houses, which were the edge of the village proper. I cannot tell you how muddling it is. We did not know which was our front. We did not know whether our own troops had come round us on the flanks, or whether they had stopped behind and were firing into us. And besides, a lot of German snipers were left in the houses we had come through, and every now and then bullets came singing by from God knows where.

Four of us were talking and laughing in the road, when about a dozen bullets came with a whistle. We all dived for the nearest door, and fell over each other, yelling with laughter, into a very dirty outhouse. James Leckie, the Old Old Man, said 'I have a bullet through my best Sandon twillette breeches.' We looked, and he had. It had gone clean through. He did not tell us till two days afterwards that it had gone through him too.

The presence of German snipers continued to pose problems, So Grenfell asked if he could try to strike back:

The day before yesterday, they gave me leave. Only after great difficulty. They told me to take a section with me, and I said I would sooner cut my throat and have done with it. So they let me go alone. Off I crawled through sodden clay and trenches, going about a yard a minute, and listening and looking as I thought it was not possible to look and listen. I went out to the right of our lines, where the 10th were, and where the Germans were nearest. I took about thirty minutes to do thirty yards; then I saw the Hun trench, and I waited there a long time, but could see or hear nothing. It was about ten yards from me.

Then I heard some Germans talking, and saw one put his head up over some bushes, about ten yards behind the trench. I could not get a shot at him; I was too low down, and of course I could not get up. So I crawled on again very slowly to the parapet of their trench. It was very exciting. I was not sure that there might not have been someone there, or a little further along the trench. I peered through their loop-hole and saw nobody in the trench. Then the German behind put his head up again. He was laughing and talking. I saw his teeth glistening against my foresight, and I pulled the trigger very slowly. He just grunted, and crumpled up.

The others got up and whispered to each other. I do not know which were most frightened, them or me. I think there were four or five of them. They could not trace the shot; I was flat behind their parapet and hidden. I just had the nerve not to move a muscle and stay there. My heart was fairly hammering. They did not come forward, and I could not see them, as they were behind some bushes and trees, so I crept back inch by inch.

As the next day dawned, Julian Grenfell returned to the German trench:

> And found it empty again. Then a single German came through the woods towards the trench. I saw him fifty yards off. He was coming along upright and careless, making a great noise. I heard him before I saw him. I let him get within twenty-five yards, and shot him in the heart. He never made a sound. Nothing for ten minutes, and then there was a noise and talking, and a lot of them came along, through the wood behind the trench about forty yards from me … They halted in front, and I picked out the one I thought was the officer, or sergeant. He stood facing the other way, and I had a steady shot at him behind the shoulders. He went down, and that was all I saw.

First World War snipers usually occupied positions in the trenches at the outset, where, among the problems to be faced and lessons to be learned, was vulnerability to counter-sniping. This placed a premium on camouflaging loopholes and finding ways to deceive enemy marksmen into revealing their positions. Frederick Crum observed in *Memoirs of a Rifleman Scout* that on one specific occasion he and his men had:

> most of their loopholes located and many of their steel plates pierced by our steel bullets. It is sad, the whole thing, but then I never looked upon it as anything else …
>
> With war once declared between the [opposing] snipers at Blangy we set to and brought all our past experience into play in real earnest. A good system of observation was established with a central post, connected by telephone, in the roof of a high red brick building, the shell of which was still left standing.
>
> Hidden, camouflaged loopholes were so constructed that every part of the Hun lines could be kept under constant and close observation. Some of these posts were so near to Fritz that the work of building them took many nights. To excavate the stones, the snipers, in some cases, dug them out laboriously with penknives to avoid being heard. The slightest sound drew bombs and rifle grenades and every change had to be gradual so as not to attract attention.

When defence gave way to attack as the balance of power changed, snipers often had to adapt. No longer could they take unlimited time over each shot, when the necessity to rapidly engage multiple targets or repulse raiders arose. Consequently, sniper rifles were sometimes discarded in favour of submachine guns in the Second World War.

In most battlegrounds, the environment changed beyond recognition as units surged back and forth: once-verdant countryside gave way to shattered urban settings, and open beaches, where there was virtually no cover, led to jungle, in which virtually nothing was easily seen. Though some courses included the exploitation of cover, even to the extent of building dummy villages, training was often too basic. Snipers had to face new challenges, and, as casualties testify, pay the price. Reality meant adapting sniping techniques to the environment, Frederick Crum remembering:

> One particular point of vantage from which we 'straffed' the Hun. We called it the 'Boiler House' for it contained the remains of the boilers of some huge ruined brewery. It was a tricky place to approach, and we had two men killed the first week. We had two bombers stationed at this point, where, peeping through a small hole in the wall, they watch for the Hun not 15 yards away.
>
> One day, going to this post with Riches, I discovered that, by creeping forward like a cat and pushing a small ladder up on a platform above the bombing post, it was possible to look right down on the Hun, and occasionally catch a glimpse of the sentry. As I was looking my attention was attracted to earth being shovelled up at a point in the trench about 50 yards off. Presently the shovelling ceased, and a healthy young Hun emerged in his shirtsleeves, spade in hand, and smoking a cigar. He rested and leaned against the wall of the trench, looking up and staring in my direction. Fifty yards through a good Zeiss glass brings men pretty near. It was hard to keep still and realise he did not see me. I showed him to Riches who, trembling like an eager terrier, was all for a shot.
>
> But I wanted to learn more. As I was watching, presently from a new direction six more Huns, carrying hods on their backs and large blocks of concrete, came up to this spot in file. I let them proceed till they all vanished with their loads down into the hole from which originally the thrown-up earth had attracted attention. What were they doing? Soon they came out and stood in a clump. This was asking too much! Riches fired and claimed to hit two, but the cloud of brick dust obscured my view … I remember clearly the expression of terror on the faces of those of them who did escape, crouching and running past us.

Klavdiya Panteleyeva reflected on the horror of her baptism of fire in 1944:

> We attacked. Then we got pinned down by that machine gun and the sniper. The regiment chief of staff Aleksey Kitaev was next to me. He had a cap with a bright band. They shot him from the beginning. He got blue, fell down.
>
> We had been warned that before shooting we were supposed to pull out all wounded. I crawled to one wounded soldier, he had a stomach wound. I started picking him up, but his intestines immediately fell out. I didn't know what to do with them, so I said: 'I'll go bring a medic'. And crawled to another wounded, because I couldn't do anything with that one. And it was so hot! He was already getting black …
>
> Fedorova was wounded, Irina Gracheva was wounded – many girls, I already forgot their names. Few of us remained. Marusiya Gulyakina was wounded again. I was shell shocked, but I didn't go to the medics, because there was blood all around, my uniform was punctured in many places, I was deaf. I thought: 'Why would I go, what can they do for me? There are people without legs, without arms, blood covered, why would I go?' And I didn't go!

Not everyone shared Julian Grenfell's view of 'Boche hunting' as a macabre sport, and, though most snipers could justify their role in some way, there were many differing opinions. On 4 July 1915, Lieutenant Oliver Hogue of the AIF (a contributor to newspapers using the pseudonym 'Trooper Bluegum'), who had been in the Gallipoli trenches when a Turkish sniper 'shot a young chap dead, clean through the head', wrote that:

> An observer saw where the sniper was located, about 375 yards off. So I grabbed the dead man's rifle and had a shot at the Turk. The observer 'spotted' for me. The sniper returned to the attack. We had shot for shot. I could hear his shots zipping into my sandbag and my observer could see my bullets kicking up the dust all round the Turk's loophole.
>
> There was a bit of a breeze and though I could not get within inches of the bull's eye, it was not until my fifteenth shot that I got right inside the loophole and silenced the enemy. He never fired again. So I think I got him. At any rate my observer said I did, so I am justified in claiming my first Turk.
>
> When I looked at the poor young chap lying beside me in the trench with the bullet-hole in his forehead, I could not feel the slightest compunction for having killed a man.

Writing a half-century later, Denis Edwards opined in *The Devil's Own Luck* that there was:

> a time to be born, a time to live and a time to die. When one of our close mates was killed we just had to believe that his time was up, and then to get on with the job for which we had been so rigorously trained. Our job was simply to fight an enemy who threatened our homeland, and, hopefully, to avenge the deaths of our comrades, by killing a few more of the enemy before our own time was up and it was our turn to die!

Yet emotions ran high: some snipers never got over what they had seen and done. For example, Sepp Allerberger had taken his friend Balduin Moser as observer for the mission. But things did not go well:

> A second buzzing bang followed the first instantly, and there was a sound like a handclap next to Sepp. Blood and scraps of tissue splashed the left side of his face. He turned to Balduin and found himself looking into a demonic grimace. The Russian marksman's bullet had been deflected by Balduin's binoculars and had exploded directly in his mouth. It had torn away his lips, incisors, chin and half of his tongue. With panic-filled eyes he stared at Sepp, while a weird gurgle of foaming blood came from his shredded oral cavity. Seconds later another bullet exploded into the earth between them.
>
> Immediately Sepp scrambled back deep beneath the tank, pulling Balduin along by his feet. Leaving their position before nightfall was impossible, because it would have meant certain death at the hands of the enemy marksman. So they were doomed to wait. Sepp felt helpless, unable to do anything for his comrade's horrible wound. No bandage or pressure dressing would help here – his only hope was fast and professional help by trained medical personnel. But this was not within reach. Sepp could only watch as the remnant of Balduin's tongue swelled to the size of a child's ball, gradually blocking his respiratory tract. Sepp tried to press the swollen tissue to the side, but Balduin started to puke and got even less air. Only a tube or a cut in the windpipe could have saved him.

Yet when Allerberger encountered Cossacks who had got inside the German positions, knowing it was difficult to shoot the riders in near-darkness, he had to aim at their horses:

> If the bullet hit the breastbone the horse fell immediately, overturning and often trapping its rider. If the hit was in the intestines or kidney area the animal started to buck furiously, became uncontrollable and finally broke down and died slowly, beating with its legs violently and convulsively …
>
> Sepp shot into the breastbone of the foremost animals, and into the soft parts of those that were further away. His comrades then shot at the downed riders. In this way they managed to fend off several attacks. After an hour the approaches were filled with dying animals, innocent victims of the excesses of human violence. An increasing revulsion of being forced to shoot these poor animals arose within Sepp.

It was almost impossible to predict how individual human beings would react in such terrible circumstances. And sometimes there were unforeseen revelations. *Sniper on the Eastern Front* tells of an incident in which Allerberger's unit were opposed by Soviet snipers and the Germans had placed dummy heads in different positions to help trace the direction of shots which:

> indicated that the Russians had a good view of the Germans, so they must have elevated positions, logically among the thick treetops. But Sepp could hardly imagine that experienced marksmen could make such a cardinal error, shooting from a tree without any possibility of withdrawing safely or taking cover. He gave the arranged signal and his comrades showed the artificial heads. Suddenly there were several shots from the Russian side and he saw branches move from the pressure waves of their fire …
>
> Sepp placed five machine guns in well-hidden positions with a good field of fire in the direction of the forest, and relocated the soldiers in charge of the fake heads. Slightly to one side he chose a well-camouflaged position for himself.
>
> Then he signalled to the riflemen to raise the artificial heads from cover while he observed the Russian line. When a head was shot, he could make out the marksman's position. Then while the machine guns fired into the treetop Sepp put in his aimed shots. This way he could camouflage his fire for as long as possible and conceal the presence of a German marksman from the Russians. The fact that the Soviet snipers had shot for their enemies' heads five times but were sitting in the trees proved to Sepp that they were good shots, but tactically inexperienced. This reduced his fear of the upcoming duel against their superior numbers.

> His plan worked with almost frightening efficiency. When a 'head' appeared, one, two or sometimes even three Russians shot at it at the same time. Sepp located the moving branches, took aim, waited for the MG salvo, fired, and hit one target after another. The hit Russians fell from the trees like sacks. A quick change of positions, and the game started all over again. Within an hour he had shot eighteen enemy snipers this way. Then suddenly the fire on the artificial heads stopped.

Allowing an hour to pass without incident, the German sergeant decided to advance into the woods while Allerberger and a machine-gun team provided cover. When it was obvious that the Russians had withdrawn:

> The sergeant waved over to Sepp and the others. Carefully, mistrusting the deceptive silence, he approached the forest and saw young women lying dead in the grass before him ...
>
> It was the first time the riflemen present had knowingly fought against women. Now that they stood before the lifeless bodies and saw the broken young faces they all experienced a strange feeling of revulsion and shame, even though they knew that in the end there was no way to avoid the law of kill or be killed. If they had known that their enemies were girls they might have fought with less vehemence and in consequence have become victims of their own morality.

Problems of a different kind arose when casualties mounted among the snipers. Herbert McBride observed in *A Rifleman Went to War* that:

> It became more and more noticeable as the Somme fight progressed and our replacements became very numerous. Our older men – that is, those members of the original Battalion who had received long and thorough instruction in rifle firing over the ranges of Canada and England, remembered and profited by their range training. These old-timers would always endeavour to take up a good position before firing, and then they really looked about for a target to shoot at. But the replacements, who had had little or no range training, would drop into the nearest hole, poke the rifle muzzle over the edge, and let 'er go. These chaps were always the first to run out of ammunition too ...
>
> During the offensive stages of an action the rifleman (sniper) does his best work alone. One man can slip through a lot of holes where a squad would be instantly detected, and it is the squads, the groups,

that interest the enemy machine gunners. They want a target worthy of the expenditure of a hundred or more bullets, not a single man.

And it is while they have their attention concentrated upon some likely target – that is, a group – that the 'lone hunter' has his chance to steal a march on them.

Seeking the Target

In many ways, the approach to a potential target away from the confines of a static or defensive position, such as a trench, was one of the most important aspects of 'offensive sniping'. A decision had to be taken what to take. The rifle, optical sight and spotting telescope were essential, together with a water bottle and a few basic rations.

The best ammunition that could be obtained was essential, cartridges being taken from the same manufacturing batch to ensure consistency of performance. Ball cartridges were used unless the task called for armour-piercing, incendiary or occasionally even tracer rounds. The sniper would have checked each cartridge visually, cleaned and, if possible, cycled it 'dry' through the rifle-action to reduce the likelihood of jamming. The weights of bullets varied – ball and tracer particularly – generating differing muzzle velocities. Consequently, the firer had to be aware of divergent flight paths, which were of real significance should the target be far away.

Atmospheric conditions could be important: the angle and position of the sun; the strength and direction of the wind; lingering morning mist over a lake or river; a chill which could highlight a tell-tale wisp of propellant smoke; changes of temperature if observation was prolonged from day into night, or if the approach was to be made at night to strike the next day.

Terrain had to be analysed, particularly if cover was minimal or natural features inhibited progress. The approach had to be mapped in the minds of a sniper and his observer to reduce the chances of being spotted. It was obviously far easier to move forward in scrubland or wooded terrain than, for example, was so often the case in the deserts or barren hills. But it was also much more difficult to identify enemy positions in woodland.

Identifying a shooting-site before embarking on the approach, if it could be done, was beneficial. And, most importantly, the locations of enemy pits, posts and trenches, threats of discovery or retaliation, had to be considered. Only when the team was confident of reaching its goal would an advance be made slowly, steadily, keeping watch, and taking detours to avoid detection.

Once 'on site', the sniper team had to ensure that it was sufficiently well dug-in below the enemy's sightline, helped if possible by vegetation and foliage. The men had to be aware that the muzzle blast that accompanied each shot could kick up the dust or snow, which could be immediate (and possibly fatal) betrayal of their presence. In *With British Snipers to the Reich*, Clifford Shore remarked that 'instruction was given in the selection of fire positions', and:

> emphasis laid on the fact that the atmosphere in the early morning tended to make muzzle smoke hang about and that muzzle blast tends to result in dust flying, or disturbance. (It is interesting to recall that during the course at Sylt a demonstration of sand disturbance by muzzle blast was staged, and after three rounds had been fired it was possible to pick up the contour like swirls of sand in front of the muzzle with the telescope at appreciable distances.)

There are many reports of snipers being killed when reflections from their rifles, equipment or, particularly, telescope-sight lenses flashed a giveaway in the sun. This posed a problem during both world wars, simply because the metalwork of the guns was customarily blued; only when sand-blasting, Parkerising and similar non-reflective finishes appeared did the problem recede. Varnished woodwork caused similar problems, though many snipers, using whatever was to hand, applied rudimentary camouflage to disrupt the outlines – though even the best camouflaged stock is worthless if no attention has been paid to obscuring the rest of the rifle!

Once the sniper team was in position, unless circumstances dictated otherwise, another period of observation began. A scouting telescope, with better light-gathering qualities than the rifle telescope and generally a wider field of view, was an excellent tool.

If the position of the enemy was unknown, the greater the value of observation became. It was not unusual for snipers to remain motionless for long periods – if they had taken position at night, for example, and were waiting for sunrise to blind the enemy.

High-status kills

Snipers were often instructed to eliminate 'high-status' targets in the hope that killing a senior officer would destabilise command structure. They were taught to recognise regimental badges, rank identifiers, and the medals and decorations that were often worn by the arrogant or the unwary. One of the most important tasks, said Klavdiya Panteleyeva:

> was to eliminate commanders, machine gun emplacements, messengers that would be running around. They also had to be eliminated. Soldiers were not necessary, mostly – officers, commanders. You would fire one shot, let go of the rifle, and lie there. You would wait until your partner fired her shot. When it became dark, we left our position. During the day we walked around, looked for a good spot to lie in wait. Sometimes picked a spot in front of our trenches. After picking a spot, took up the position when it was dark. Then we lay there without moving a muscle until the next evening, because you couldn't crawl away in the daylight. If there was an attack, that was different, then you would get up and run. Otherwise, you would lie in that spot to the end.

A clue to rank could be detected if men were giving commands, carrying binoculars, or had handguns instead of rifles. German snipers had killed many British officers in the First World War simply because they wore jodhpurs or knee-high puttees, or carried swagger-sticks. Clifford Shore reflected in *With British Snipers to the Reich* that:

> The dress and equipment worn or carried by an officer or NCO should so clearly resemble that of the men that they should be indistinguishable through a telescope at 500 yards. But these excellent policies were not carried out to the letter by some units. Until the 'officer and NCO indistinguishable from the men' policy had been adopted there were many leader casualties. In Italy during the battle for Carroceto a forward company of one of the Battalions in a Brigade had the Company commander, a subaltern and every sergeant killed by German snipers. These snipers worked well forward of their line and reserved their shooting for targets wearing field glasses, a soft cap, or carrying a map instead of a rifle or carbine. After this, orders were issued in many units that officers would not carry binoculars slung round the neck, and the majority of officers proceeded to carry rifles.

Russian snipers killed many German officers during the Great Patriotic War simply because rank distinctions and Iron Crosses were still to be seen. In an extreme case, a single well-placed shot could turn the tide of battle. Vasiliy Zaytsev recalled in *Notes of a Russian Sniper* how:

> A hunched-over German soldier approached the enemy trenches. He was unarmed, and only carried a bucket in his hand. We decided

to leave him be, since he looked so shabby and pitiful. Another ten minutes went by, when suddenly a heavyset, polished Nazi officer turned the corner of one of the trenches. He had a colonel's insignia on his jacket. A sniper followed behind him, carrying a beautiful hunting rifle with a huge scope.

Two additional officers emerged from behind the same corner of the shallow trench. One of them was a major wearing a Knight's Cross with oak leaf clusters. Following behind the major was another colonel, who was smoking a cigarette in a long holder ...

Our shots rang out, three volleys of two shots each. We had made textbook head shots, and all four of the Nazis dropped to the earth, their lives draining out of them.

Specific targeting

Individual snipers often favoured particular types of hit to sap the enemy's morale. Lyudmila Pavlichenko deliberately shot the second man in a column; done often enough, this made the 'second man' in any line fearful. In extreme circumstances, no one would 'go second' and confusion reigned until officers and NCOs gave the orders, which could result in near-mutiny. Other snipers deliberately shot targets in the chest or stomach, particularly if they were part of a group. Colleagues would rush to the injured soldier's aid, exposing themselves to another shot, and medical staff would be needed to deal with the wounded in a way that a 'clean kill' could never do.

Shooting the last man in a column, especially if strung-out or detached from the group, could pass unnoticed and allow another shot. Firing by precise time, 'death at eight o'clock', could instil ever-growing fear as the clock ticked down (though never to be attempted without constantly changing shooting position).

Observing supply tracks, shooting messengers or those attempting to reinforce forward positions could also be profitable. Said Evgeniy Nikolaev:

> The purpose of a sniper is to cause disorganisation in the enemy's direction of its forces by destroying its manpower through precise shooting and to strike first and foremost at the most important targets, which may appear just for a short time: command personnel, look-outs, artillery spotters, scouts, signallers, ordnance, mortar and machine gun crews, anti-tank teams, motor cyclists, parachutists, but, first and foremost, enemy snipers. By directing fire at armoured targets, observation slits and periscopes, our snipers can blind and destroy tank crews and fire at descending enemy aircraft.

> Thus, for example, in the autumn of 1941 a soldier from Tambov, Yuri Semyonov, a sniper of the 602nd rifle regiment in the 109th Internal Affairs Ministry rifle division (Leningrad front), hit a pilot with a single shot and downed an enemy plane, for which he was awarded the Order of Lenin. And this is not the only case in sniper operations.

There were many instances of snipers firing to such good effect that large numbers of enemy soldiers proved unable to fulfil their tasks. For example, John Fulcher recollected that he had spotted German troops:

> at the end of the road where it hazed into the horizon. I nudged my partner and nodded in their direction. We continued to squint into the sun as the troops became individual soldiers marching in company formation, like they were in a Berlin parade for the Fuehrer or something. Through binoculars, I could tell they were green replacements. Their uniforms were still a crisp gray green; their jackboots kicking up little spurts of dust still shone.
>
> They left a cloud of dust hanging in their wake. Apparently, they were on their way to the front. I looked at my partner. He had his rifle scope trained on them. He looked back at me. He shook his head.
>
> A whole company?
>
> Peering through my rifle scope, I could see the oily glisten of sweat on the commander's brow. Most of the time you could tell the officers. They wore sidearms and were always shouting and waving their arms.
>
> I cross-haired the officer and shot him through the belly. He looked momentarily surprised. He plopped down on his butt in the middle of the road. The report of the shot reached him as he fell over onto his back. He was dead by the time I brought my rifle down out of recoil and picked him up again in my scope. His legs were drumming on the road, but he was dead. His body just didn't know it yet …
>
> The other krauts were so green they didn't know enough to scatter for cover until my partner got in his licks by knocking down one more. Even then, they behaved more like quail than combat troops. They hid in the drainage ditches and in some shell craters, their heads bobbing up. Just like quail. I figured I could have drilled two or three more, but I held my fire. It wouldn't do to be pinpointed, even by green troops.

An uncharacteristic lack of response from the Germans duly awakened John Fulcher's curiosity:

> When there were no more shots, a junior officer in the ditch rose cautiously to one knee. Shading his eyes, he stared at the ridge in the opposite direction from where we were hiding. He gradually gained confidence when still nothing happened. He stood up and looked around. He carried a Mauser. He walked out into the road and stood a moment and looked at the two dead men. That seemed to do something to him. Turning suddenly, he ran back to the ditch and started flapping his arms and haranguing his cowering soldiers. Finally, the fearful replacements clambered to their feet.
>
> The company reorganized without making an attempt to find us. It tore down the road again toward the front.

There were other ways to disrupt an enemy's routine – hitting metal loophole plates in the dead of night to ruin sleep, or shooting holes in water containers to exploit thirst. Vasiliy Zaytsev recalled one effective incident:

> Instead of singing birds, morning was announced by the spitting of machine guns and the whistle of rounds flying overhead. We had set up our sniping position at the very front edge of our lines. Our orders were to take out the Fritz officers, then their NCOs, and finally to use grenades on their equipment …
>
> As expected, an uneven fire-fight broke out. Our only possible escape routes were cut off – retreat was impossible. We dug in for a siege.
>
> Twenty-four hours passed, then another twenty-four. We managed to hold out. With our sniper rifles we could control the access to the spring, and pretty soon the Fritzes were getting parched. We made sure to expedite this by shooting holes in all their jerry cans.
>
> We thought it was an excellent joke, and we listened to the Fritzes cursing and whining as their water drained away. 'Pretty soon they'll be drinking their own piss', said Nikolay Kulikov …
>
> 'We have to keep them from reaching clean water', I instructed the other snipers. 'We want to make their guts run.'
>
> So our sniper group forced the Fritzes to give up their excursions to the spring, and we were also able to turn most of their mortar crew into casualties.

The tools for the job

Of course, missions were aborted, even after prolonged observation, simply because the sniper's target was impossible to define with certainty. It was far better to try another day than take a chance. The unwise fired and risked retaliation; the experienced sniper could accept defeat. Or they could ensure that something more effective dealt with the threat.

During the early days of the First World War, for example, British snipers sometimes used large-bore 'elephant rifles' to penetrate the iron shields that supposedly conferred immunity on German snipers. In the later conflict, the Russians often used 14.5mm anti-tank rifles to neutralise snipers' posts which were either out of range or too well protected to attack with 7.62mm sniper rifles. Serving as a sniper with the Sixth Airborne Division, Denis Edwards describes in *The Devil's Own Luck* how on 15 June 1944 – near the Château St Come in Normandy – a fairly quiet day had:

> started with the normal dawn stonk of between twenty and thirty mortar bombs and shells. Later, we took turns in going out to the front on patrols and sniper hunts. One team decided upon an ingenious innovation – they took a PIAT with them. When a sniper was spotted a PIAT bomb was fired at the top of his tree. We had no way of measuring the success of their effort other than from the fact that several of us thought that fewer snipers were bothering us after this incident.

And Adrian Gilbert, in *Stalk and Kill*, describes how one Australian soldier and his partner had used a .55 inch Boys anti-tank rifle to bring down a Japanese sniper:

> With the barrel resting on my shoulder, the butt against his own, Harry took a long aim, quite undeterred by the bursts of bullets from all sides. Then Harry fired and I was crushed to the ground and Harry was flung against a tree and the sniper toppled gracelessly out from behind his tree.

Raking treetops with machine-gun fire became a popular way of dealing with arboreal snipers. In addition, there were even occasions when the support of field guns, tanks or even aircraft was arranged.

Acclimatisation

Once snipers were in position, there was range to the target to consider. But many factors can influence range-gauging. Underestimates can be due to clear weather, if the sun is behind the firer; if the terrain is rough or undulating; when shots are taken uphill; when the lower part of the target is obscured; if the shot is taken across water or snow; or if the target contrasts too sharply with the surroundings.

Conversely, overestimates may occur if it is rainy or cloudy; at dawn, dusk or in shade; when the shot has to be taken against the sun; or when shooting downhill. Shots will fly high into a headwind, hit low in a tailwind. And it is notoriously difficult to make accurate assessments when prone.

Snipers could generally rely on nothing but their own observational skills, though attempts had occasionally been made to provide ranging stadia and similar marks on the reticles of telescope sights. Success depended on the skill with which they could make allowances for range, wind and light. Some zeroed their sights at a specific distance; others adjusted their sights in advance for each change in distance. Neither method was infallible, even though they saved experienced snipers the vital seconds taken to adjust the sights in combat – a life saver, giving time for the near-instant second shot that would otherwise have been impossible.

Mikhail Kondratyuk, after noting that he kept eating carrots in the belief that they contained vitamins that were good for eyesight, continued:

> Some think you just take aim, press the trigger, and goodbye, Nazi! No, you need to be more sophisticated. Excuse me for the question, but could you take out a Nazi at a range of eight hundred metres? Can you imagine the science required? Let me tell you. First, you need to determine that he's eight hundred metres away, and not six hundred or seven hundred. For this you need a refined eye. And to work out distances using angles, you need to know geometry.
>
> When a bullet is fired, it spins from left to right and deviates to the right. At six hundred metres the deviation is twelve centimetres, and at eight hundred it's as much as twenty nine. Knowing this figure, you train your sights accordingly. But if there's a strong cross wind, what then? You have to move your aiming point by a couple of digits. But there may be a variety of circumstances. There could be a wind, or the Fritz could be running – and all in different directions …
>
> There's so much addition and subtraction, it's enough to make your head burst. And you've only got three seconds. Even a professor would be hard pressed.

Kondratyuk related how his duel with a German sniper had led to helpful information:

> I was sitting in a hide-out on the Mius river. I was after the Fritzes on the other side. But it had been more of an embarrassment than a hunt. For three days I had not potted a single one of them! Shameful! So, I

> tested the rifle's range again, ate half a kilo of carrots, and consulted with the captain. All in vain – the bullets were falling short, so at night I swam naked across the river with a rope to check the distance. That didn't help. Then I wrote to sniper Chekulayev. And – what do you think? – got a telegram back: 'Over water obstacle allow bigger vertical angle, because cold air and damp lower trajectory'.

Reaching out

When the distance to the target exceeded that at which a successful hit could be guaranteed by a single rifleman, experienced snipers sometimes fired as a group, setting their sights to bracket the distance estimated to the target.

This greatly increased the chance of a hit, even though some sights would be set too low while others were too high. Clifford Shore wrote about a:

> conversation between the four snipers … as to the range to the German observer [in a tree-top]; they could not agree. Two of them said they estimated the range to be between 250 and 300 yards; the others said it was nearer 350 yards. The matter was settled in a rather novel manner. Three of the snipers set their sights at 250, 300 and 350 whilst the fourth sniper took the binoculars and kept them riveted on the prospective target. When quite satisfied … he coolly gave the three men a fire order; the three rifles 'spoke as one' and the Bosche came somersaulting to the ground.

Sometimes, circumstances dictated that shots should be taken at extreme range. For example, USMC Private Daniel Cass Jr described an attempt to neutralise hostile Japanese machine-gun fire, made on Okinawa in 1945, with the assistance of Corporal Chester Carter acting as observer:

> We found our way to the top of a ridge overlooking the valley, all the time listening to the intermittent chatter of the Nambu as it grew louder the closer we got. Mud collected on our feet like twenty-pound weights. Fog clinging heavily to the peaks slithered into the lowlands and writhed in thin veils along stream beds among the battered remains of forests … The Nambu let loose a string. I ducked instinctively then looked again. I noticed sweat ran in rivulets from Carter's lean face. I felt myself sweating too.
>
> It was like having the front row seat to a movie. The voices of the Marines on the floor of the valley sounded thin and reedy as they were whipped into shreds by the rapid *Clack! Clack! Clack!* of the Japs'

> machine guns. Marines scurried among a field of shell craters and shattered forest; they sprawled motionless behind whatever cover they could find. A few of them lay in the open. They were dead. Nips high on the ridge opposite from us sprayed them with a deadly hail of lead …
>
> At least 1,200 yards of valley separated us from the machine gun nests. Fog made for such poor visibility that I could not tell where the firing came from. Desperately, I used my rifle scope to search for gun smoke or some movements to give the Japs away. Carter used his spotting scope. I saw caves and coral ledges and ragged stumps.
>
> Down in the valley a Marine jumped up to improve his position. An invisible finger fell out of the air and seemed to flip off his head. The body went tumbling and flopping. That lent further desperation to our mission.

Carter then reported that he had found the Japanese machine guns, allowing Cass to follow his point to:

> a spot located just below a coral ledge honeycombed with shallow caves. The nest was well camouflaged and dug in, but I detected a thin screen of smoke coming from what appeared to be toppled underbrush. The smoke was a little bluer than that made by shellfire, and it was darker than the shredded remnants of fog. There was no wind; the gun smoke clung to the air above the boulders and felled trees. Pinpoint muzzle flame flickered from the shadows. Two, maybe three, guns were rattling from that location.

Carter estimated range to be 1,200–1,300 yards. This forced Cass to consider the variables that:

> entered into shooting at a range of even one thousand yards, the longest shot I had ever tried. Wind and heat waves and, in this case, fog distorted the scope picture. The ammo we used was not match grade; it came right out of the green boxes. Sometimes you got misfires with it; sometimes a round simply went poof! and you had no idea which way it went.
>
> I looked about frantically. The ridge we were on cut away from the valley the farther you went down it, increasing the range across. I saw I had no choice. The Japs were chewing up the Marines below us …
>
> After some quick calculations, spurred on by hopelessness, I slipped a tracer round into the chamber of the Springfield and indicated to

> Carter that I was going to use the telephone pole above the enemy placements as an aiming stake to try to get the range.

Looking through his telescope, Carter saw the strike point and advised Cass to raise it by an inch. So he:

> adjusted the scope elevation knob and dropped cross hairs onto the Japanese machine gun nest. The Nips had to expose themselves at least a little in order to pivot their gun barrels down toward the valley floor. I heard Carter murmur something … as I narrowed my concentration against a tiny patch of grayish uniform exposed above a tree trunk. I tried to remember everything I had learned at Camp Pendleton about long-distance shooting. So much depended on it. Smoke curled up from behind the logs as the Nambu commenced to chatter its short repetitive phrases.
>
> 'Deep breath. Let half out. Hold. Cross-hair, cross-hair, squeeze.' With the spotter scope, you could actually pick up the streak of the bullet going through the air. Carter grunted. 'All right', he said when my first round plunged into the enemy's barricade.
>
> I worked the bolt with a feeling of elation. My hands and breathing were surprisingly steady; I fired and worked the bolt, fired and worked the bolt, pouring accurate fire into the Japanese defenses, cross-hairing handkerchief-sized targets momentarily exposed more than a half mile away.
>
> Even through his spotting scope, Carter couldn't tell when I scored because the targets were so fleeting they disappeared whether I made a hit or not, but he was all grins …
>
> The machine gun fire ceased … That was when I took a deep, wavering breath. I had done my job, hadn't I?

When it came to fire, snipers had to take control of their breathing and be aware almost of their heartbeat before gently squeezing the trigger. They also had to heed the moment: it was vital to fire at the instant that a target was exposed to its greatest extent, but also to fire before an opponent recognised any threat.

Take the risk

An ability to improvise effectively has always been one of the sniper's greatest assets:

> One morning several medium-sized German tanks suddenly broke through from a gully to a village on the steppe ... Not far from Goncharov lay a young Uzbek anti-tank gunner, seriously wounded in the stomach by a German shell splinter ... The sufferings of the wounded man lit a fire in Goncharov's heart. Without uttering a word, he picked up the anti-tank rifle, which had fallen from the hands of the dying Uzbek, pressed the butt against his shoulder, put his eye to the sights and, without even feeling the beating of his own heart, pressed the trigger. Leaving a smoky trail, three bullets, one after another, sped like lightning over the seared steppe, and everyone saw three initially small puffs of smoke tenaciously clinging to the enemy vehicle. Within a few minutes they had combined into one big black puff. Then the tank erupted in a firework display.

This incident led directly to Pyotr Goncharov's career as a sniper. He described a subsequent battle with two enemy strongpoints:

> I saw them firing their rifles, and bullets were pinging past my ears, but I couldn't pick up their damned pill-boxes. Then they would go quiet, but I knew that as soon as our lads got to their feet, the enemy bullets would cause hundreds of deaths again. One of our troops couldn't restrain himself, dashed forward, and was cut down in a second.
>
> Suddenly I caught sight of a small mound. I took a closer look, and saw it was a pill-box! It was about four hundred metres away, no more. Well, I thought, now we can get even! I had a good sight of two Nazis through the telescopic sights of my rifle, just like a mirror. One of them was firing, while the other was holding the ammunition belt by the machine gun. I didn't even bother to check the distance to the pill-box with a tracer bullet, as I usually do, but fired three bullets, one after another, at the gun-port. And straight away the machine gun fell silent, as if somebody had cut it out of the picture. I let them have it with anti-tank bullets!
>
> When the lads looked later, only splinters were left of their Nazi skulls. Then, the battalion charged forward, and things became easier. But another pill-box started firing. This one was a bit further away, about 600 metres. But time was precious, and the lives of my comrades were even more precious. I had pre-loaded my rifle again with armour-piercing cartridges and blazed away again at the gun-port. I emptied the whole magazine without pausing for breath. So, that one fell silent too, and I still couldn't believe it. I lay there and just couldn't believe that I had stuffed its throat with lead!

Gennadiy Velichko took part in a counter-attack made during the battle for Stalingrad. Nikolay Filippov related in 'Ognennoe chuvstvo' ('Fiery Feeling'), published in 1971 in *Krasnaya Zvezda*, how Velichko had noticed:

> A thin column of blue smoke rising up in a gully behind the trenches. It came from a German field kitchen …
>
> Soldiers began to appear with their mess tins, making their way towards the kitchen. Soon a whole throng had assembled. It looked as if an argument had broken out between the cook and the soldier nearest to him. Both of them were vigorously flapping their arms. It was all a bit far away, but Velichko fired. One of the soldiers slumped down and immediately sprawled on the ground. But nobody in the crowd paid any attention. Velichko took a second shot. And a third. And each one hit its target. The soldier who had been arguing got his due and slumped like a heavy sack by the hooves of the horses. He was followed by another, who fell face first to the ground. The crowd scattered in all directions. Velichko also managed to get the cook.
>
> Steam was pouring out of the field kitchen. He was not going to leave the Nazis' dinner for them! The sniper loaded his magazine with armour-piercing cartridges and fired. A thin stream shot out of the resultant bullet hole. The horses reared up. The field kitchen turned over.

Getting away with it

A shot was only really successful if the sniper, observer and associated personnel could withdraw safely to their own lines. Consequently, not only was it essential to plan the approach but it was equally important to plot the escape in such a way that the retreating team could not be seen. This could be time-consuming, but time had no significance if safety was to be ensured. Wise snipers sometimes planned several routes, selecting the best route as circumstances dictated.

Successful kills usually had to be validated by officers to add to individual scores. But intelligence gathered by the sniper team, which could monitor enemy movements for hours at a time, was often more useful. Alterations in the enemy's lines, changes in personnel or the presence of reinforcements, sometimes the prelude to an attack, could influence tactics.

Virtually all sniper instructors and sniping textbooks have emphasised the importance of reporting observations accurately. This was, therefore, inevitably a keystone of training.

Filippov's *Krasnaya Zvezda* article related that Second Lieutenant Golosov had noticed through his binoculars that 'someone was crawling with a submachine gun along an enemy trench' about 400 metres away:

> Velichko set himself up on the parapet and began to take aim … The figure twitched, and everyone saw his helmet fly from his head. At Gennadiy's firing point there were three trenches connected by a communication passage. He sat in one of the trenches and sought out further targets in front with his predator's eyes. An enemy signaller appeared and started to mend the line that had been broken by our mortars. Velichko potted him with his first bullet. But he also gave himself away.
>
> A Nazi sniper opened fire on him. Gennadiy hung his helmet on a stick in the trench, tied a piece of twine to it, raised it a little, then moved into another trench and from there began to tug the string to give the impression that someone was moving around in the first trench. The sniper fell for it. His bullet struck the helmet, the helmet fell down, and the sniper took a quick look out of his hiding place to check on his hit. It was at this point that he parted company with the world of the living.

And Aleksandra Medvedeva related how:

> Every day and night snipers went to the front line – for a 'hunt'. They dug in and observed the relocations of Germans on the front line. This was the place I opened my sniper's score.
>
> I remember how I killed my first fascist. Together with my partner Zina Vershinina we occupied our sniping positions. While observing enemy positions, I spotted a machine-gunner. I aimed and shot. It was unclear whether I hit him or not. But when I returned to the detachment, everyone already knew I'd killed an enemy. An artillery observer reported this. He saw it in his periscope, how the enemy machine-gunner was killed. Everyone was joyous, they hurried to congratulate me. And I wept, for I had to kill a man …
>
> I – a common girl – had a hard time getting accustomed to the front and shoot at people, although I understood I was shooting at the enemy. Soon there was a crisis in my conscience. Seeing people's sorrow, tears and blood of my native land, I understood there could be no pity to the fierce invaders. For the killing of my 10 first Hitlerites I was awarded with the medal 'For Courage'.

The most memorable of her many experiences of combat, she said, happened when:

> In the Autumn of 1944 there was heavy fighting in Poland. An enemy sniper appeared in one location, and we suffered appreciable losses: our commander, a scout and many signalmen were killed. There came an order: 'Destroy fascist sniper!'
>
> Three of us went to carry out the order: Nina Isaeva, Lena Akulova and I. We settled upon our positions and started shooting at roofs with armour-piercing incendiary bullets. Thatched roofs caught fire fast, the Germans started escaping those houses with their mortars and machine guns. And we shot on and on … We were so carried away with this, that we completely forgot about staying cautious. And the German sniper noticed us. He hit Nina Isaeva right in the eye by a well-aimed shot. Luckily, she stayed alive, although she lost her eye and her vision was significantly impaired.
>
> Having concentrated ourselves, we still discovered the lair of the enemy sniper and eliminated him, as well as other firing positions of the Germans. Our task was fully accomplished! For this successful battle operation I was awarded with the Order of Glory, 3rd Class. The war went on. I had 43 eliminated fascists on my personal account. I was awarded with the Order of Glory – 2nd Class.

Antonina Kotlyarova opined that:

> Studies are one thing, they can teach you a lot at school, but when we came to the front, everything was different. Of course, first thing we did was to look into the embrasure to see what the German was doing. And on the very first day a girl from Leningrad was killed. She spent the entire blockade in Leningrad, and on the first day after our arrival, when she looked through the embrasure, a bullet ricocheted and hit her right under the eye. So our wartime service began with a funeral.
>
> When I killed my first Fritz, I returned, a reporter approached me and wanted to interview me. I don't know what I told him, but on that day, and the next one, I couldn't eat or sleep. I knew he was a fascist, that they had attacked our country, they killed, burned, hanged our people, but he was still a human being. It was such a condition …
>
> When I killed the second one, I was in a horrible state again. Why? Because I saw him through my optical sight: a young officer. He seemed

> to look at me and suddenly I killed him. But he was a human being! Basically, a horrible state. But afterwards I grew somehow desensitized.

Nina Lobkovskaya, whose personal tally had reached 89 kills by May 1945, recalled an incident that had happened:

> in the Kalinin Region. My partner and I were summoned to a place where our soldiers were shot daily. No one could tell where the shooting came from. So we started observation. We spent a whole week examining the area in every detail. We would take up position before sunrise, so as not to be spotted by the enemy. And one day, quite unexpectedly, I looked into my sniper eyepiece and saw a German in a white shirt, a high-collar jacket draped over his shoulders and a peaked cap.
>
> I was shocked because he was so close! I sized up the distance and aimed. It was a hasty shot, so the bullet only smashed his cap badge. He understood that the shooting came from a sniper, shook his fist at me and disappeared behind his disguise.
>
> I remember him well enough: a young man, rather good-looking, I'd say. I couldn't sleep that night, my mind was racing as I was trying to recall every detail, every trifle. From then on, a hunting game began, each stalking the other. I felt his presence, and he felt mine, I could tell that. The duel lasted for a week. One day I got careless and let him spot me. He fired right away. The bullet hit the metal of the back-sight, ricocheted into the helmet and singed my temple. As I saw blood, I thought of one thing only: to fire back. So I snatched my partner's rifle and began to watch him from another angle. I knew he was bound to reveal himself somehow. And he did, for I soon saw the line of his helmet above the bushes. I took aim below that line and shot. No more shots were fired from the spot …
>
> I was satisfied, but I will remember his hand clenched into a fist shaking at me for the rest of my life.

Eventually, as the Red Army surged westwards into German territory, the fighting drew to a close. Klavdiya Panteleyeva recalled that when her unit reached the Baltic:

> There was an offensive at Koenigsberg, we captured it. Then we were in the defense. We didn't participate in combat there, only in the defense. A sniper in general was supposed to be only in the defense. At the end of the war, they didn't take us anymore.

> Then the war ended ... While we were in the trenches we weren't doing anything. They were feeding us. The prisoners marched for several days. I don't know now how many days that was. Then they transferred us to a forest, there was discipline there. We had nothing to do, so they had us build paths in the forest, with little borders, so we would be busy the entire day.

Kill the Sniper

The rise of sniping inevitably encountered reprisals. Though these could take many forms, the 'enemy sniper' was the villain who had to be eliminated almost without regard to the cost. Writing in *The Australian Guerrilla*, Ion Idriess recollected that:

> We clustered like bees around the lip of the well. A bullet smashed a man's temple and the blood sprayed across the clear water surface. Soon we got into quite a hot spot – a series of stumpy sand mounds each with its clumps of shrubby bush. Turks waited behind every mound ... We rode in among hills of rolling sand, bare of bushes which lessened our fear of snipers.
>
> Three chaps jumped off their horses and peered among the gloomy trees, when 'crack!' Corporal Logie slung up his rifle and 'crack! crack! crack!' sang through the palms. One (Turkish) chap was kneeling behind a palm frantically waving a tunic ... the other was down ... the poor fellow was shot through the arm and clean through the body, fatally. They were snipers, their ammunition belts were half empty ... We breathed freer when we climbed to the circle top of grim hills enclosing that nameless oasis. To have been caught down there would have meant a death trap.

Among the most successful counter-snipers have been Vasiliy Golosov of the Red Army's 81st Guards Rifle Regiment (70 snipers in a total of 422 kills gained by the end of July 1943), Lyudmila Pavlichenko of the 54th Rifle Regiment (36 in 309), Zhambyl Tulaev of the 580th Rifle Regiment ('thirty or more' in 313), and Yefim Kiryanov of the 81st Guards Rifle Regiment (30 in 200, until September 1944).

Publication of the memoirs of Vasiliy Zaytsev and Lyudmila Pavlichenko, particularly, have done much to publicise counter-sniping. Zaytsev's duel with ace German sniper 'Major Konings, commandant of the Berlin Sniper School' has been related many times, and was a central feature of *Enemy at the Gates*.

But Zaytsev's antagonist remains shadowy: at the time of writing, no convincing evidence of his existence has ever been revealed. Though there can be little doubt that experienced marksmen would have been entrusted with neutralising Zaytsev, the story reeks of propaganda. The same can be said of the duel between Lyudmila Pavlichenko and Helmut Bommel, related in *Lady Death*, though the story told in her memoirs differs considerably from those published in wartime Soviet patriotic journals.

However, almost all of the leading snipers engaged in counter-sniping as part of their routine. Zaytsev himself tells of an encounter in Stalingrad, beginning with detailed preparation, which sought to eliminate a particularly troublesome German marksman who had killed several Russian snipers:

> Nikolay Kulikov was waiting for me, at our position opposite the pile of German shell cases. We had stashed an artillery periscope there, to aid us in surveillance. What we saw when we looked through it didn't surprise us: twenty-two shell cases lay where they were the day before. The twenty-third, with the missing bottom, had disappeared. But where had it gone to?
>
> An artillery periscope has excellent optics. Small details of the terrain, or of a soldier's uniform, will be clearly visible from several hundred metres away. I ran my eyes over bush after bush, foxhole after foxhole, using the magnifying optics to penetrate deep into the enemy's lines. I scanned to the very summit of the hill. Next to the summit was a small depression, and there it was – the artillery shell with the missing bottom! It was camouflaged by an embankment, near the hollow. Through the shell case, I could see the German's scope. The spent case shaded his scope from sunlight, hiding its reflection, and helping its operator see his targets. It even seemed that he had some camera apparatus on his rifle, and was photographing each target, at the moment he fired his shots.
>
> 'There you are …' I whispered, as if he would hear me and depart on his own. I was shaking with rage. In order to calm down, I passed the periscope to Nikolay Kulikov and told him where to find the sniper. I walked his gaze there, guiding him by easily recognizable landmarks.
>
> 'Got him! There he is!' Nikolay exclaimed. 'Chief, we've got to nail him now. Otherwise, he may get away.' …
>
> By now I had held the Fritz sniper in my sights for over an hour. I was tired and losing my mental focus. I forced myself to banish this exhaustion. Meanwhile the Fritz kept peering through his scope. It was only a matter of time before he discovered us. I asked Kulikov, 'Nikolay what do you think – does he see us?' 'We'll find out'.

Kulikov moved back to a position where he could raise the helmet a few inches above the embankment on the end of a stick:

> The German fired a shot that ripped through the helmet. I was surprised that he went for this bait. Perhaps the tedium of waiting for a target had made him forget the risk he was taking.
>
> I watched as the German sniper placed one hand on the breech and reached forward to pick up the empty cartridge case. Collecting spent cartridge cases was standard operating procedure after making a victorious shot. As he did so, he raised his head slightly from his scope. It gave me the few inches of scalp I needed to zero in on … and at that second my own shot rang out. The bullet struck him at the hairline; his helmet fell forward over his brow and his rifle lay motionless, the barrel still inside the shell case.
>
> Leytenant Fedosov dropped to the trench floor and found my little note pad that held my 'personal headcount'. Fedosov licked the lead of his pencil and inscribed in large letters: 'I have witnessed a duel. Before my eyes, Vasiliy Zaytsev killed a Nazi sniper. – A. Fedosov'.

Typical of the activities of Eastern Front counter-snipers were events in the careers of Feodosiy Smolyachkov, regarded as one of the major inspirations behind the rise of Soviet sniping, and Moisey Usik:

> Smolyachkov was stalking the enemy sniper for the second day running. It was the first time he had encountered such a cunning opponent. It was clear that the sniper had come to this sector with a definite objective. But what was it?
>
> Plenty of snow fell at night and a frost suddenly descended towards morning. The bright sun illuminated the plain, the young copse not far away, and the hollow where Feodosy was concealing himself. The fresh snow hung thickly on the branches of the trees and bushes. Weighed down, they were bending towards the ground.
>
> Without giving himself away, Smolyachkov kept watch on his opponent's position. How many times he felt the urge to press the trigger and take out an idle enemy soldier! As if to spite him, traffic that day was more lively than usual. But Feodosy steadfastly resisted the temptation. He had to neutralize the Nazi sniper, whatever it took. The main thing was not to give himself away.
>
> Smolyachkov's interest was drawn by a small hillock which rose slightly above the snow. Something glittered beside it and immediately

vanished. There was somebody there! Smolyachkov carefully pulled his rifle towards him. Whereupon a bullet whizzed past above his ear.

It was the sniper … so he was there, by the hillock. The lens of telescopic sights glinted. Feodosy bent his head, pressed himself more firmly into the snow and again grasped his rifle. A second shot resounded and another bullet screamed past above his head. The situation was becoming more difficult. He had to think of something. But what?

Smolyachkov returned to watching for his adversary:

About fifteen metres away Smolyachkov noticed an old fat tree stump – a solid defensive object. It would be good to get over there, but how? He couldn't make a dash. But if he crawled under the snow … The idea was alluring. Without delaying for a second, he began to burrow under the snow, raking it away in front of him. Fortunately, the snow cover was quite thick. These fifteen metres which he traversed in order to reach the tree stump seemed to Smolyachkov to be a good dozen kilometres long.

The duel between the Soviet and the Nazi snipers had already lasted over two hours. The sun began to set. Reddish tinges appeared on the snow and the trees … Feodosy noticed that part of a foot was visible by the hillock behind which the enemy was hiding and immediately fired. The Nazi sniper's foot twitched, and for a moment the side of his body was exposed. This moment was enough for Smolyachkov to finish his opponent off with another shot.

For Moisey Usik, it had been a difficult night:

Flares flew up ceaselessly, whether from our own or the German side, illuminating the front line with an unreliable light.

With the soldier assigned to assist him Usik crept up to the broken-down tractor. Paying no attention to the bullets whistling by in the dark, the soldiers began to dig a fox-hole, trying not to crush the tall weeds around the tractor. They dug in turn, one wielding the spade while the other lay shovelling the soil into a bag and scattering it among the shell holes. At the same time, they had to keep watch constantly to check there were no Nazi scouts nearby.

It was a dangerous move that Usik had undertaken. One slip and he could give himself away. But Usik reasoned that they could not shoot

> from the loft any more, the German sniper could not be spotted from a defensive position and, consequently, they had to find a place from which they could take him out. In this respect a fox-hole under the broken-down tractor was the most convenient ploy.
>
> The fact that it was close to the German trenches made it even better – at least for the time being it would not occur to the Nazis that a Soviet sniper had dared to position himself right under their noses. He just had to hit the target with the first bullet so as not to give himself away and, of course, to camouflage himself well.

Usik's hiding place proved to be very good, protected by the weight and bulk of the Stalin tractor with its wheels firmly dug into the ground. After making a peephole in a corner behind the tractor, camouflaging it with dock leaves, the sniper could hear 'brusquely voiced cursing in German, which sounded like barking dogs … from the enemy trenches'. Through a chink between the tractor axle and the ground surface, he had:

> a good view of the regional hospital with its gaping window apertures and of the left flank of the enemy defensive forward line, which stretched back to the ruins of the barracks of the former infectious diseases block … Usik thrust his rifle between the scraps of iron and began to view the German flank through his telescopic sights. At that moment a tousled German head next to a pile of bricks showed up in his sights. From the angle of our own defences he was concealed by a parapet, but from here he was visible till half way down his chest. Usik was seized by a big temptation to fell the Fritz, but prudence won out; the main thing was to wipe out their sniper.

However, time passed. Sunrise lit up the hospital, but still Usik could see nothing to target. Heat from the mid-day sun was making the fox-hole beneath the tractor uncomfortable when he noticed that in:

> the frame of the second window from the edge in the second storey of the hospital a bright column of light appeared in the depths of the building and then vanished. A second later the column appeared again and just as quickly disappeared. It resulted from thin streams of dust and refuse being illuminated by the sun's rays shining through the gap in the wall and alternately glinting and fading out.
>
> 'Why is the dust falling? There doesn't seem to be any shooting or wind. There must be somebody walking round there', Usik quickly

> concluded. And, as if to confirm this, a dim shadow appeared on the opposite wall and disappeared again.
>
> Several seconds passed. The shadow did not reappear. But then a brick fell down. The muzzle of a rifle appeared in a gap between the bricks. Then some semi-circular reddish object began slowly rising up over a gap in the wall.
>
> 'He's a cunning one. Rubbed his helmet with powdered brick. And I didn't even think of covering my own with dust. But it's all right. It's not visible under the tractor'.

The German was becoming increasingly confident, standing at full height as he laid cartridges on a board by his side:

> As if carrying out the orders of his own nemesis, [the German] compliantly appeared in full stature at the edge of the gap in the wall. He was now aiming in the direction of yesterday's house, apparently taking the dummy for a Soviet sniper.
>
> Usik did not hear the shot. Rather he felt it instinctively because at that moment he also fired himself. Hitting the edge of the wall and bouncing up, the rifle fell from the window.
>
> Scraping the building stones with his hands, the Nazi fell heavily from behind his cover and lay across the window.

Evgeniy Nikolaev, facing hardships 'when it was already winter', related in *Red Army Sniper* how he had:

> fought a real duel with a German sniper; he was an ace shot, but I got the better of him. I spent three days tracking him down, got close to his firing point and waited for him, lying under a metre of snow … Dressed in a summer uniform in a thirty degree frost, with an overcoat, a forage cap on my head, artificial leather kirza boots and not a single glove.

Ilya Grigorev killed at least 18 German snipers during his career. One such duel, settled by trickery, had been with a German known as 'Hans', who 'killed with impunity Soviet troops and officers who carelessly looked out from their trenches, and stalked our signallers'. Grigorev was operating:

> a little to the right of the Nazi sniper, He seemingly vexed the Nazis even more than Hans vexed us. One night the Nazis announced

through a loud-speaker: 'Russky, remove your Ilya and we will remove our sniper Hans'.

The enemy found out about Grigorev from leaflets which ended up in their hands and described the sniper's heroic deeds. On hearing the Nazis' 'request', Grigorev went to see the regimental commander and asked permission to enter into a duel with Hans. After thinking it over the commander replied curtly: 'All right!'

Grigorev chose what seemed to him to be a suitable site, marked out some reserve spots, and began to keep watch. He noticed that the bushes on one small rise occasionally quivered in an unnatural way. This put Grigorev on his guard. He saw something flash up there and then a shot resounded. In an instant a bullet rapped against Grigorev's helmet, which was covered with camouflage netting.

Realising that he had fallen for a trick, Grigorev changed position and:

pulled from his pocket a stick which he had prepared earlier, tied a mirror and a long cord to it, and crawled to his reserve position. Then he tugged the string and the mirror flashed. He tugged it again, and suddenly a shot rang out! The stick fell down, and the mirror shattered into small pieces. But at least the Nazi had now given himself away. Grigorev tied a shard of the mirror to a new stick and, taking aim at the point where the spotted, square helmet emerged, he pulled the string. Almost simultaneously two shots rang out.

The Nazis did not have the opportunity to recall their Hans; his body lay in no-man's land for a long time, while Ilya Grigorev continued to inspire fear among the enemy and strike them unerringly.

Sometimes even the most experienced sniper could fall for a simple trick. John Fulcher used one straight out of the Hollywood playbook:

A shot echoed ahead in the ruins. From where I was, all I could see was the backs of GIs as they ducked behind parts of walls and buildings, crouching there and not moving …

A few minutes later, I heard another shot. And another. I figured it for a sniper, but I held my men tight until a runner came sliding in next to me. He was panting with excitement. 'The old man wants snipers. That fucking kraut's got three of us drilled up there already'.

I tapped the sniper next to me for a spotter. The runner stayed behind with his head down while my spotter and I dodged forward

> from building to building until we reached the commander crouching behind the remains of a house. He was breathing heavily, and sweat had matted plaster dust on his face until he resembled a mime.
>
> 'We don't know where the hell the bastard is out there', the CO said.
>
> I had to chance a look. I crawled to the end of the building. I had learned to memorize every detail of terrain after one quick peek. I took that peek from right next to the ground, past some kind of bottle half-filled with spoiled food.
>
> To my front was a street that ran away from me to where it intersected another street filled with debris from a recent shelling. A bed with the mattress still on it rested atop a pile of concrete wreckage. One of the dead GIs lay at the intersection. He seemed so much a part of the general waste that you hardly noticed him. By his position, I figured the sniper's hide to be farther down the street where he could control both it and the intersection.
>
> 'Keep everybody down and in position', I instructed the CO. 'Give us a little while. We'll see if we can take care of him'.

The two soldiers worked carefully forward, cutting across an alley out of the sniper's view and along a 'shell-pocked wall overgrown with dead ivy' to a point where they could settle. Fulcher related:

> 'Krauts don't watch Roy Rogers,' I whispered to my partner. 'Let's give him a target and see if he shows himself'. We had used the trick before.
>
> My spotter crawled down to the end of the wall nearest the street while I eased my M-1 through a crack clogged with enough ivy to camouflage my movements. As I watched the terrain through my scope, my spotter put his helmet on a stick and slowly thrust it above the wall. He jerked it back down and waited a minute or two before revealing it again. This time the sniper was ready.
>
> Krauts didn't watch Roy Rogers or they would have recognized the oldest sagebrush trick in the Old West.
>
> An 8-millimeter German bullet drilled the helmet and sent it spinning.
>
> I caught movement through my scope. A thin puff of smoke popped from the window of a building about a hundred yards down the street. There wasn't much left of the building, but it still had enough roof over its remaining window to hold shadows inside.

> I held my fire. It was like when you spooked a squirrel into its hole. All you had to do was sit down and wait. Sooner or later curiosity overcame caution and the squirrel stuck its head out again. I kept my scope centered on the window, waiting.
>
> A good sniper, a survivor, never shot more than two or three times from the same hide. I was counting on this German being careless or lazy.

Though initially unable to see inside the house clearly, Fulcher then detected a slight movement:

> I shifted my cross hairs slightly and bided my time. Hunting animals gives you the patience to hunt man. Soon, I picked up the dull momentary gleam of a rifle barrel. Then a faint outline materialized as the kraut peeked out to see what was going on. Just like a squirrel.
>
> I fired. An invisible rope yanked the German's weapon from his hands and hurled it into the street. The German's body lunged forward across the window sill where it hung head down, his hands fluttering a moment against the street. The body spasmed a few more times and drained a dark pool of blood below its head. Then it was still.

But even this trick didn't always work. Clifford Shore, in *With British Snipers to the Reich*, described what had happened in 1943 on the Lamone river during the Italian campaign:

> The British and the Bosche each had control of their respective banks. There was a considerable amount of rifle shooting backwards and forwards across the river, some of it sniping and some not! Our snipers kept a very careful watch and one day enjoyed themselves hugely for a few seconds.
>
> On this day the Bosche playfully placed some steel helmets on sticks and proceeded to walk along the river bank sticking up the steel helmets so that they were just in view over the top of the bank, and apparently with the German's characteristic guttural chuckle awaited for our snipers to waste ammunition and give away their positions by firing at the steel helmets. But the Huns concerned lacked foresight, and forgot that there might be snipers at the bends [in the river] – and there were! – and that they might be enfiladed. For a good percentage of the mischievous Huns it was a last prank.

Reflecting in his memoirs on a highly successful service career, Bruno Sutkus noted that he had fought:

> fifty-two sniper duels and survived many of them only because the opponent misjudged the range and so fired too short or too high or failed to allow correctly for the wind. I am sure I had a sixth sense that enabled me to spot enemy snipers very rapidly.
>
> My weapon was a Zf-K98k … with the low turret sight fitment and Zeiss Ajack 4x telescopic sight. Ammunition was important. I had special ammunition: the best was captured Yugoslavian and that of German pre-war manufacture, which gave tighter groups.
>
> The sniper must always be aware of the position of the sun and take care that it never reflects from his field-glasses and so betrays his position. I had a special camouflage jacket issued only to snipers. It was earth-brown with a white lining and could be worn inside out if needed. It was hooded, of ample cut and was therefore comfortable in all shooting positions.

It has been suggested that as many as half of all snipers who saw combat were either killed or wounded badly enough to leave the line (if only temporarily). Among the most unusual case is that of Iosif Pilyushin, author of *Red Sniper on the Eastern Front* (2010), whose service career in and around Leningrad seemed to have been terminated by the loss of his right eye. Amazingly, Pilyushin then taught himself to shoot left-handed and was able to resume his combat career. But there were unexpected hurdles to overcome:

> I began to train vigorously: I dragged cases with cartridges and grenades to the front and hauled beams for constructing pillboxes and bunkers. Each morning I spent time crawling on my belly and leaping ditches. When leaping I couldn't always judge the distance correctly [with one eye] and I often didn't reach the other side of the ditch …
>
> [Eventually] I took a lot of shots and successfully so: the shots were all on target, although the bullet holes were scattered. Despite all the difficulties of firing with my left eye, I had achieved my main objective: I could defend myself in combat. With each shot, the bullet pattern in the target became tighter and tighter. But it still required a lot to calculate each shot accurately from any distance.

Tallies: Fact or Fiction?

Many of the snipers active during the First World War have been credited with specific numbers of kills, most notably the Canadians, though Billy Sing and other Australians are usually treated in the same way. But it is no longer clear how the tallies were assessed. Though Sing's commanders were prepared to support his claims, it seems that reliance was widely placed on the veracity of the individuals' own recollections

Comparatively little information is available concerning British snipers in particular, largely owing to the low esteem in which the contributions made by men such as John Fairall (killed by a sniper's bullet in 1915) and Alexander 'Sniper Sandy' Macdonald, whose tally is said to have reached 97 when he was killed in November 1916, were regarded once peace had returned in 1918.

The same can be said of performances by men such as Harry Furness (117 kills?) in the Second World War. There can be little doubt that British snipers made notable contributions to individual campaigns, yet most of them remain anonymous. Clifford Shore observed in *With British Snipers to the Reich* that:

> In Italy, one private soldier had over 60 certified kills to his own rifle. A South African officer had more than forty kills and got the Military Cross for his sniping, but was eventually killed himself.

But Shore names neither man: his book was published in 1948, at a time when the individual deeds of war, still raw, often passed unacknowledged. The same is true of his comment about the performance of the Australians fighting the Japanese in Timor, where:

> One of these kangaroo hunters [now tentatively identified as Private Douglas Weatley] had a great sporting time, and whether right or wrong, he played sportingly with the Japs and never used his telescopic sight … [at] less than 300 yards range! He was credited with 47 Japanese killed but with characteristic modesty claimed only 25 certainties …
>
> Another 'roo hunter presented with a mass target of automaton Japs – the target which comes once only in the life of one sniper in a hundred – got twelve of the yellow-men in twelve shots in fifteen minutes.

In addition, few individual claims were ever verified. Writing in *The English Review* in 1920, Eric Penberthy said that among the many volunteers and impressments were a few who sent in:

> glowing reports as to the number of 'kills' they made every day. As there were few means of checking such statements, these had either to be accepted, without proof, as genuine or disbelieved altogether. In one battalion a sniper I knew generally headed his daily report something like this:– 'Sir, I have the honour to report that six little wooden crosses will be required in the Hun trenches this evening, making a total of 11 this week. I have the honour to be, Sir. Your obedient servant, No.___, Pte.___, etc., Sniper.'
>
> As the Hun snipers on this front persisted just as strongly in spite of this man's supposed heavy execution, the Brigadier finally sent a message to the C.O. saying: 'In future the left ears of all Huns killed by Pte.___ will be attached to his reports, please'.

By way of confirmation, Trooper Allan Campbell reflected that at Gallipoli:

> Some Anzac snipers confidently claimed big tallies of their shooting. My experience is that the sniper could never be sure of killing his enemy. Only if a man is stooping when shot through the head will he spring. In any other position when shot, he collapses unseen by the shooter. It is unusual when sniping to have target upon more than the enemy's head. So beware of false reports.

By the end of the Second World War, the Russians, and to a lesser extent the Germans, had snipers with almost unbelievably large tallies. Mikhail Surkov of the Red Army's 4th Rifle Division claimed to have registered 702 kills. Then came Vladimir Salbiev of the 71st and 95th Guards Rifle Divisions (601 to 31 December 1944), Vasiliy Kvachantiradze of the 259th Rifle Regiment (534), Akhat Akhmetyanov of the 260th Rifle Regiment (502 to 15 January 1944) and Ivan Sidorenko of the 1122nd Rifle Regiment (500).

Snipers were often pressed into service as assault troops, gaining no credit for kills achieved with rifles or submachine guns in this role. Consequently, it has been suggested that at least one man, Fyodor Okhlopkov, serving the 1243rd, 234th and 259th Rifle Regiments, reached 1,000 kills if his performance as a submachine-gunner is added to his sniper-score of 429.

Of course, killing a high-ranking commander could have a far greater effect on the battlefield than the deaths of 100 low-ranking men. The kills attributed to Litva Rugova – 242 or 265 depending on source – are said to have included 118 officers; she recalled in 1980 that: 'We were told that killing its leaders would destroy the German army, so that's what we did.'

Many commentators in Britain and the USA dismissed the Russian snipers' claims as fabrications, and even Andrey Ulanov, writing in *Stalkers of the Enemy – Red Army Snipers*, quotes research undertaken by historian Igor Chernyaiev who, while assessing records of the 257th Independent Rifle Brigade, had:

> discovered a report about a sniper company from the 7th motorised rifle division of the NKVD internal forces during the period August 21st–24th. The total NKVD tally totalled 642 enemy soldiers and officers. However, according to German data, the German 296th infantry division which happened to be operating on this sector lost only 227 troops dead and wounded.

And it is probable that scores were compromised by the difficulties of confirming results: for example, a blow to a helmet which knocked the wearer backward could also give the appearance of a kill. Klavdiya Panteleyeva acknowledged in 2010 that it was often impossible to tell if a kill was actually a wound; success, she said, was usually acknowledged if the target 'fell down' when struck.

Yet Russian snipers were supposedly only credited with kills if there were witnesses. In theory, each success had to be verified at first by a colleague and then, after the Order of Glory had been introduced in November 1943, by their commander. There are many testimonies, including that of Vasiliy Zaytsev:

> Before I managed to say a word, Sasha Gryazev had picked up two grenades and darted off down the trench. 'Stop, get back here!' I yelled.
>
> Sasha halted and turned to me with a real hang-dog look. 'Chief, permission requested to kill the enemy. I'll take out the entire position with one go', he said with a chuckle. Then he continued, in a more serious vein. 'Chief, they're not all for you to kill. Let the rest of us chalk up a few.'
>
> This rebuke caught me by surprise, but I should have anticipated it. The rest of the guys had been keeping track of their kills – I wasn't the only one. We'd set up daily 'personal head counts'. Each sniper had to confirm his kill with an eye-witness's signature.
>
> It was true, I had quite a few more than anyone else. Sasha Gryazev's own signature attested to this. So, what could I say to him?

But Zaytsev admitted that discrepancies sometimes arose between:

> the kill figures reported in dispatches and those in my own personal record book. Observers would sometimes inflate my kills, because they would simply count how many shots I had fired. To them, three shots meant writing a 'three' for the number of kills. Of course, those observers could not see my targets as well as I could, so they never knew any better. Whether or not my bullets actually penetrated their intended targets – only I knew this.
>
> This sort of 'score-keeping' saddled the snipers with a tremendous responsibility. It was the law of mutual trust coming into effect. We had faith in the monitors' numbers; and they took us at our word when we said we had hit a target.

Sergeant Bill Jones of the US Army said of his solitary authenticated success that 'confirmed':

> meant you had to go up and look at his teeth shining between his lips and pick out the bullet hole. Most of the time when I zapped one, the other Germans either dragged him off or we were too fagged out to worry about going up to take a look at where he fell. All dead men looked the same anyhow.

Asked in 2010, Klavdiya Panteleyeva confirmed that the trench commander had to sign the sniper's record, and then the sniper returned with evidence to present to his or her superiors.

But in trench warfare or the shattered industrial landscape of Stalingrad, for example, the results of each and every shot were not always easily observed. Consequently, scores are quite likely to be overstated owing to the difficulties of confirming a kill.

There are many reports of snipers being hit, not fatally but in circumstances which could easily lead the firer to claim success. Nina Isaeva, for example, survived a bullet that hit her in the right eye. Yet even if 10 per cent is subtracted from each sniper's total, there can be no doubt that the Soviet snipers were not only exceptionally effective but also set the standards to which others aspired.

The performance of the German snipers, excepting several well-known men, remains largely unknown. The highest scorer was Matthäus Hetzenauer, with 346 confirmed kills, though the career of Bruno Sutkus – supported by a diary in which countersignatories are identified – is far better documented.

This is particularly true of the period before the introduction on 20 August 1944 of the *Scharfschützenabzeichen* (marksman's badge). There were three grades of these cloth badges, bearing an eagle's head within a wreath of laurel

– 1. *Stufe* for 20 kills, with plain oversewn edging; 2. *Stufe* for 40 kills, with an edging of silver braid; and 3. *Stufe* for 60 kills, with gold braid.

Each success had to be verified by an accredited witness (officer or NCO). Kills obtained prior to 31 August 1944 went uncounted, recognised only by the award of an Iron Cross. Snipers of the Waffen-SS, and later of the Ersatzheer and Volksgrenadiere, could gain additional honours. Bruno Sutkus suggests they ranged from five days' leave for five kills to recommendation for an award of the German Cross in Gold for 100 kills.

Men whose tally reached 150 were to appear personally before Heinrich Himmler, Reichführer-SS, to receive a suitably engraved watch or be invited on a hunting trip.

In Their Honour ...

Accredited Soviet snipers wore a badge introduced on 21 May 1942, and were usually awarded the medal 'For Courage' when they had achieved 10–12 kills. After the approval of the Order of Glory on 8 November 1943, however, kills were only accepted if they could be verified. The performance of snipers prior to this date, therefore, can be difficult to authenticate.

Even in times of great conflict, excepting in the Soviet Union, the contributions made by snipers – often including self-sacrifice – were all too rarely recognised. This was due in no small part to the public view of sniping, which in its simplest form was seen as little more than a necessary evil and at worst as officially sanctioned murder. Consequently, few men were ever recommended for the highest awards. Only two Victoria Crosses awarded during the First World War are said to have gone to snipers: posthumously to Private Thomas Barratt of the South Staffordshire Regiment, killed near Ypres on 27 July 1917, and to Sergeant Albert Lowerson of the 21st Battalion of Australian Infantry Force for bravery at Mont St Quentin in September 1918.

A citation published on 6 September 1917 in the *Supplement to the London Gazette*, no. 30272, notes that Barratt, born in 1895, serving the 7th Battalion, South Staffordshire Regiment, gained his VC for:

> Most conspicuous bravery when as Scout to a patrol he worked his way towards the enemy line with the greatest gallantry and determination, in spite of continuous fire from hostile snipers at close range.
>
> These snipers he stalked and killed. Later his patrol was similarly held up, and again he disposed of the snipers.

> When during the subsequent withdrawal of the patrol it was observed that a party of the enemy were endeavouring to outflank them, Pte. Barratt at once volunteered to cover the retirement, and this he succeeded in accomplishing.
>
> His accurate shooting caused many casualties to the enemy, and prevented their advance.
>
> Throughout the enterprise he was under heavy machine gun and rifle fire, and his splendid example of coolness and daring was beyond all praise.
>
> After safely regaining our lines, this very gallant soldier was killed by a shell.

Lowerson had joined the Australian Imperial Force in July 1915. Too late for the Gallipoli campaign, he saw combat on the Western Front on the Somme, at Pozières and Mouquet Farm. Recovering from wounds, Alby Lowerson rejoined the 21st Battalion in November 1917, but was wounded again during the Battle of Mont St Quentin, returned to his unit in September 1918, was wounded yet again and finally invalided out of service.

There seems little doubt that he had acted as a sniper at some point in his career, but the citation for his Victoria Cross, published on 13 December 1918 in the *Supplement to the London Gazette* no. 31067, does not specifically refer to such a role.

Anonymity often complicates the identification of snipers, particularly if they were serving with infantry units. As casualties were disproportionately high, many men took their successes to the grave unrecognised. It has been suggested, though evidence is lacking, that 80 per cent of CEF snipers who served during the First World War were killed or wounded, and that British Army sniper casualties during 1918 were around 50 per cent.

Yet examining the *London Gazette* reveals many sniper-related awards. On 16 January 1916, for example, the *Supplement* recorded that a Distinguished Conduct Medal (DCM) had been awarded to Serjeant William George Blakemore of 1st Battalion, Scottish Rifles:

> For consistent gallantry between 3rd October and 14th November, during which period he was instrumental in removing a large number of the enemy's snipers by his excellent marksmanship.

And, according to the *Supplement* of 1 April 1916, a similar award was made to Lance-Corporal David Webb of 1st Battalion, Cheshire Regiment:

> For conspicuous gallantry and resource on many occasions whilst scouting in obtaining valuable information. Near Ypres he reconnoitred the German positions, located a hostile gun, and killed a sniper who had been doing much damage.

The Russian leadership saw the Red Army's snipers as grist to the propaganda mill, exploiting high-scorers such as Zaytsev and Pavlichenko alongside the more glamorous Roza Shanina. Occasionally, special events were created – such as the celebratory snipers' rally held in Beketovka, near Stalingrad. At the 64th Army's headquarters, General Mikhail Shumilov said:

> It's a long way to Berlin. We have no choice but to traverse it and punish those who brought war to us, raised the sword against us. There must be no rest for the enemy day or night; we must destroy him without mercy! You must be good and attentive teachers to new detachments of sharp-shooters. Everything that you have accumulated during the great battle on the Volga must be passed on to your pupils.

Shumilov drew attention to the 'painstaking work, tactical nous, dexterity, and inventiveness' each sniper had shown, urging them to share with one another 'ploys which had been borne out by combat experience, in order to make sure of hitting the enemy and taking him out with the first bullet'. Nikolay Ilin, one of many high-scorers present, responded:

> The Soviet government has awarded me the title of Hero of the Soviet Union. I am a sniper. In the battle for Stalingrad I wiped out two hundred and fifty eight Nazis and my full tally comes to three hundred and fifty eight. I vow to strike the enemy even more ruthlessly.

Having received a sniper's badge, Ilin returned to his seat and asked his pupil, Afanasiy Gordienko, 'What do you think?' One observer related how:

> The constant hard work, the privations, and the face-to-face contact with danger were no burden to Nikolay. He seemed to be singing an endless song of joy. The amusing jokes and wise-cracks which accompanied his every step bespoke the festive nature of his soul. And this mood was bound to be transmitted to others, so that he infected everyone with his indomitable fighting spirit. Nikolay loved to meet with young snipers from neighbouring divisions. He strove to pass on his own experience.

A similar event was described by Evgeniy Nikolaev:

> On February 6th, 1942, the entire country learned that, by an edict of the USSR Supreme Soviet, the Tambov-born senior sergeant Vladimir Pchelintsev had been awarded the title Hero of the Soviet Union. And on February 22nd Pchelintsev was invited to the first front-line rally of snipers. By that time my fellow Tambovian had wiped out a hundred and two Nazis. It was a joyous and memorable day for Vladimir. The top snipers of the Leningrad front gathered in the brightly illuminated Smolny palace. Particularly notable was the moment when Andrey Zhdanov appeared at the presidium table. He presented Vladimir Pchelintsev with the Order of Lenin and the Gold Star of Hero of the Soviet Union. Along with Pchelintsev this lofty title was awarded to Ivan Vezhlivtsev and Pyotr Golichenkov. Feodosiy Smolyachkov received it posthumously.

Assessing effectiveness of the Russian snipers is complicated by lack of information, particularly in the period before the advent of the Order of Glory (*Orden Slavy*) on 8 November 1943. The first awards were made on 28 November to junior lieutenants, warrant officers and NCOs.

Three grades were given progressively, and anyone attaining the highest level was considered 'Full Cavalier of the Order of Glory' (*Polniy Kavaler Ordena Slavy*). Nearly a million third-class Orders of Glory were awarded, but only 102 of them went to graduates of the Central Women's Sniper School.

Merely 16 of these snipers received the second-class award, including Roza Shanina, and there was only one female recipient of the first-class award: posthumously to Nadezhda Petrova. Born in 1893, 'Mama Nina' had won the All-Union shooting prize in 1939 and was eventually allowed to serve as a sniper despite facing the Red Army's customary prejudice against women. She died on 2 May 1945 in a vehicle accident, with 122 kills to her name.

Matthäus Hetzenauer, the subject of *Eastern Front Sniper*, is generally reckoned to be the most-decorated German sniper, receiving the Iron Cross first class on 25 November 1944 and the Knight's Cross of the Iron Cross on 17 April 1945. He was also awarded the German Cross in Gold, the highest bravery award made to serving soldiers, virtually as the war ended.

Evgeniy Nikolaev reflected upon a gift to Vladimir Pchelintsev:

> A presentation sniper's rifle with a metal plate gleaming on the butt and the inscription: 'To Fascist exterminator V. Pchelintsev from the Leningrad Front political administration' … This rifle struck at the

> black heart of the Nazis; the sniper's tally of vengeance rose to 152 in eight months, and among them were a number of enemy snipers, officers, look-outs and aircraft spotters.

The value placed by the Red Army on snipers sometimes extended to their rifles, which were passed on in the hope that the success of the previous owner (who had usually been killed or incapacitated) would be perpetuated by the recipient. In *Sovetskie Snayperu v Boyo 1941–1945*, the Nikolaevs related the fascinating story of Obr. 91/30 KE 1729:

> In 1940 Khusen Borezhevich Andrukhaev was called up by the army. In the same year he became a member of the Communist Party. After graduating from a military-political college in 1941 he joined the 733rd Rifle Regiment of the 136th Rifle Division and became a political instructor. Both back at the college and now in an army company he enjoyed 'having a shoot' – and improved his firing skill.
>
> When it became necessary to go into battle with the Nazis Khusen Andrukhaev took up a sniper's rifle. In a short period of time he wiped out several dozen Nazis. In the autumn of 1941 the 136th Rifle Division was involved in heavy defensive engagements against the Nazi hordes in the Don Basin. On November 8th, in the area of the village of Dyakoko … was invaded by a battalion of enemy infantry from the SS Viking division.
>
> When the company commander was seriously wounded … Andrukhaev took upon himself the responsibility of command. The company repelled three enemy attacks and the Nazis suffered big losses. But the junior political instructor was also wounded – an enemy bullet had penetrated his cheek. Having bandaged up the wound, Khusen Andrukhaev continued to … wipe out the Nazis, striking at the enemy with his sniper's rifle.

Running short of ammunition, the Russians then withdrew. Andrukhaev volunteered to cover their withdrawal, but next day, once the Germans had finally been repelled, his body was found. He had blown up both himself and several attackers with a hand grenade. On 27 March 1942, Junior Political Instructor Khusen Andrukhaev was posthumously awarded the title of Hero of the Soviet Union: one of the first snipers to be so honoured. His regimental colleagues decided to present rifle KE 1729 to their best sniper and so, on 26 November 1942, it was given to Andrukhaev's pupil Nikolay Ilin.

With his tally allegedly standing at 494, Ilin, who was also made a Hero of the Soviet Union, was killed on 4 August 1943. The rifle once again survived its owner, passing to Ilin's pupil Afanasiy Gordienko, who, in turn, proved to be an outstanding shot with more than 400 kills to his name. Retired at the end of 1943 after being damaged by the shell that killed Gordienko, KE 1729 is now owned by the Central Museum of the Armed Forces in Moscow, with a presentation plate fixed to the right side of its butt.

The Obr. 91/30 rifle presented on 12 April 1943 to Nikolay Galushkin, whose tally had reached 135 of an ultimate total of 418, was displayed for a time in the museum in Kirov. Plates were also attached to the butts of the rifles presented to the best students of both the men's and women's sniper schools

Evgeniy Nikolaev's rifle, which apparently survives in the musuem of his home city of Tambov, has a collection of silver stars on its butt. The rifle used by Rimma Uyutova, according to her own testimony, had 17 notches cut in the butt. And by December 1944 the butt of Fyodor Okhlopkov's Obr. 91/30 rifle bore 420 notches – one for each of his kills. Mikhail Kondratyuk asked:

> Have you seen my rifle? There's a silver plate on the butt, a personal gift from the industrial workers of Izhevsk. Look what's written there. Except that the tally now is not a hundred and eighty, but two hundred and sixty two. I won't update it in the meantime. Who knows what the final figure will be? No point in scraping the metal for nothing.

AFTER THE FIGHT

When the world wars ended, any reputation snipers had achieved rapidly dissipated. Indeed, many of them – even high-scorers such as Australian Billy Sing – died all but unnoticed.

This was largely due to the low status they enjoyed in peacetime, many people regarding them as little more than killers, whatever justification there may have been for their roles. Corporal Thomas Durst of the King's Royal Rifle Corps related how:

> After the last time I was wounded I put in for a commission, which meant that by the time I got through OTC [Officer Training Corps] it was October 1918 and the war was almost over. I was glad, though, I don't mind admitting it. I had been wounded three times as a sniper, and I knew it was just a matter of time … I don't think any of our sniper section came through the war, they were all killed or wounded. For years after the war I used to wake in the early hours and get up and go walking, to try and forget some of things I'd seen and done.
>
> There were lots of men like me, some of them would be walking all night but we never talked about it, though sometimes you'd get in conversation with someone and they'd ask if you'd been 'over there' and you'd say yes and you'd ask each other where you'd served, but that was about it. I never, ever told anyone I had been a sniper, not even my wife. They wouldn't have understood.

Though many German snipers thought similarly in 1918, the war for some of them was not over: many became involved with the right-wing Freikorps attempting to quell the left-wing Spartacist uprising and similar revolts which led to bloody street-fighting in towns and cities.

Friedrich Schulze, who had served 24. Infanterie-Regiment as a junior officer during the war and subsequently emigrated to the USA, recalled that:

> When the ceasefire was announced, I was teaching sniper recruits and my officer called out to us, 'You are lucky men, the war has ended!' I was so tired, I could not really understand what it meant to stop fighting. We returned to Munich in December 1918 and were immediately asked to join the Freikorps in defeating the Communists. I had learned so much in the trenches that the fighting for me was easy and we did much street sniping but shooting fellow German soldiers who had been in the trenches like me sickened me. One day I went home and threw my rifle in the river then burned my uniform and I never picked up another rifle from that day.

Accompanying her unit all the way to eastern Germany, Antonina Kotlyarova related that:

> We finished the war on the Elbe. We were staying in German houses. At 3 AM we were raised by an alarm. There was shooting, but we couldn't understand, could those be Fritzes? The end of war was announced. We ran to the basement, found some rotgut there, and drank it, chasing it down with jam.

She also recalled that she 'once got a parcel during the war':

> sent to me personally by some engine driver. I forgot now. The commander took out the alcohol, but gave me the rest. They removed vodka from all packages, and gave the rest to the girls. There was food there.
>
> Trophies? We were soldiers. No. Later, in Germany, you would enter a house – if you had to change underwear, you would open a closet, take out a slip, pants were ours, but you would change underwear. We didn't wear foot bindings anymore, but instead put on 5–6 silk or Persian thread stockings, and then boots on top. Then you would go on. I got burned on this at the end of the war. Changed … We didn't have knapsacks, we had left them with the supply platoon a long time ago. Gas masks as well. We only carried ammo, grenades, a rifle. And so, right before the end of the war, I found a house, there was clean underwear tied with ribbons in the closet. I couldn't check how it was, right?! So I changed all underwear, and it turned out to have been darned! So I returned to Moscow wearing a darned slip.

Klavdiya Panteleyeva remembered that:

> After we crossed the Niemen, we girls were walking. There were husband and wife walking toward us, they seemed old back then, but now I would've said young, about 50, they were carrying milk. So they gestured to us – drink. The girls refused, what if it was poisoned? I was uncomfortable that we were refusing, they were offering it from the bottom of their heart. So I said: 'I'll drink it'. And drank the milk, nothing happened to me. At Koenigsberg we were invited to someone, they set a nice table, treated us. Later we were invited again, ate bilberry with milk.
>
> As far as trophies are concerned: near Smolensk, you would come out at night, there would be the smell of ashes and only chimney flues were left standing. And in the daytime it was hot in July. If there were rags lying around, we would take off our boots, throw out our sweaty soiled foot bindings, pick something suitable, put them on, and be on our way. It was hard to carry what we had, we were barely walking ourselves, what trophies? Our scouts sometimes gave us chocolate. Not only gave us, once they set an entire table, there were all kinds of chocolate. They captured some stockpile of it, and invited us. So we stuffed ourselves with that chocolate.

But the thrill of victory, for many individuals, was to be short-lived. Denis Edwards reflected long after the Second World War that:

> Many have read my amplified diary notes. Many have been curious to know how I felt as I lined my rifle sights on a 'target', knowing that as I fired I was probably killing another human being. My answer has always been that as a highly trained Airborne sniper fighting in a war where you killed – and could just as easily be killed – I could not afford to regard the enemy as human beings. To me they were not boys with mothers and homes where people loved them, they were simply 'targets' to be hit as accurately as possible and they were no different from targets on the rifle range, except that they had the ability to fire back.

Obscurity and Rehabilitation

Getting back into 'normal life' proved to be very difficult. Many snipers had lost family members – Iosif Pilyushin, for example, had lost his wife and two sons in Leningrad – and others had suffered injuries, physical and mental, which were to blight the remainder of their lives. Pilyushin, who had lost one eye in combat, eventually went blind. Lyudmila Pavlichenko suffered so badly that she took to drink, dying comparatively young, and there was a widespread tendency in the USSR to characterise female combatants as 'front-line whores' when the reality was that, though consensual fraternisation undoubtedly occurred, attention had often been forced on the women by the men. Disfigurement was another problem many individuals had to face.

Some snipers found relief by compiling the notes and diaries from which memoirs have since been created, but the postwar Soviet authorities were not particularly keen to focus on details which could be read as critical of the wartime regime. This extended not only to the destruction of the museum created in Leningrad to memorialise the siege but also to the execution of the most important of its creators: the official view was that Stalingrad was to be regarded as pre-eminent in the struggle against the Germans. Eventually, however, the situation changed. Yuliya Zhukova reflected in *Girl with a Sniper Rifle* how:

> For the first time since that memorable parade on 24 June 1945, the Victory banner was borne out onto the square and among the group who carried it were Heroes of the Soviet Union Mikhail Yegorov and Meliton Kantaria, who had erected this banner on the Reichstag in Berlin. Military units and cadets of military academies, from the Suvorov and Nakhimov colleges, filed past in a triumphal march … Everything was majestic, beautiful and impressive. But back then, and when I recall that parade now, it seems to me that there was something incomplete about it, as if some element was lacking or something important omitted.

However, after a few minutes, she added:

> Diffidently at first and then more boldly, they [the attendees] began to go out onto the square, gathered in groups, shared jokes, laughed and took photos. This went on for a long time. Whatever way it came about, a feeling of festivity, joy and enthusiasm overtook everyone. A memorial medal was forged for the Victory jubilee.

So began the process of rehabilitation, recognising the contributions of individuals in a way which had been largely overlooked. Many snipers took part in interviews, historians and researchers began to look into their past, and the first memoirs were duly published. In addition, veterans began to attend reunions. Yuliya Zhukova was one of them:

> February 20th, 1975 has stuck in my memory for ever. I was sitting at my desk in my office, when I opened the newspaper *Komsomol Pravda*, looked through the first article, and turned the page. On the second page an article entitled 'sniper' immediately caught my eye. Without suspecting anything, I began to read. And suddenly … It was about our sniper school! About its graduates! I was barely able to read to the end. It turned out that a council of veterans from the school had long been active and the former cadets and instructors were meeting almost every year.
>
> With fingers trembling from excitement I … inquired how to telephone the article's author, G. Alimova. I immediately rang her and incoherently explained that I was also at the sniper school … and I would like to be put in touch with someone. She gave me the phone number of Nina Solovyei, chairwoman of the council of school veterans.
>
> I immediately rang her and found out that there would be a reunion of graduates from our school in June in honour of the thirtieth anniversary of Victory. Whereas only graduates of the first two cohorts had been invited to previous reunions, it had now been decided to involve everyone who had attended the Central Women's Sniping School. What a commotion followed!

And this was just one of many events that would be held in the USSR and then Russia – the only places in which snipers have been regularly honoured publicly for their contributions to victory.

APPENDIX 1

A TYPICAL SNIPING COURSE

Hesketh-Prichard's course was based on specific requirements, expanded when necessary to provide skills necessary to enable snipers to act as observers and intelligence-gatherers.

On **Day 1**, a general lecture on the objects of the course and discipline was followed by examination of rifles brought by students – many of which proved to be defective. Then came a lecture about the care of arms, shooting to obtain groups, and observation on a German trench.

Day 2 began with a lecture on the use and care of the stalking telescope. Then came a repetition of grouping practice at 200–300 yards, paying particular attention to bullet strikes on target. The afternoon was given over to practical observation of German trenches and in open country.

Day 3 began with a lecture on the P/14 Enfield, followed by how to judge distance up to 600 yards, snap-shooting at 100–200 yards with a four-second target exposure, application at 200 yards, and consideration of shooting positions. Map-reading and long-distance observation with reports took care of the afternoon.

A lecture on map-reading occupied the morning of **Day 4**, followed by shooting at 400–500 yards and then at unknown ranges up to 400 yards. The afternoon was devoted to use of ground and cover, selecting and constructing hasty observation posts for open warfare, and selection of cover from view rather than cover from fire.

Day 5 began with a lecture on patrolling and scouting, followed by shooting at 300 yards and then snap-shooting at 100 and 200 yards with a three-second target exposure. A demonstration of camouflage and its uses took up the afternoon, snipers being given an area in which to disguise themselves using local material. The efficacy of this would be checked by observers, who watched while the traineees fired blanks from their concealed positions and would provide map references to expose any snipers they had detected.

The principles of elevation and windage were explained during the morning of **Day 6**. Spotting enemy snipers was practised, followed by 'snap-shooting and movement' in which the students were expected to advance unseen from 500 to 100 yards while shooting at 'enemy head' targets which appeared at intervals in the butts. The afternoon was spent building and using night-firing boxes, then investigating how observation of enemy trenches could be compromised by moving materials around.

Day 7, customarily a Sunday, was one largely of rest. The range was opened so that volunteers could shoot under the supervision of an instructor. Students could take the day off if they chose to do so, but experience showed that most of them were keener to improve their marksmanship.

Day 8 started with demonstrations of the construction of forward and sniping posts, patrolling using night goggles, use of cover, and keeping in touch with colleagues. Shooting practice was undertaken at ranges unknown to the firers so that the accuracy of their estimations could be checked. A march involving the use of compass bearings, and using the box respirator from time to time to protect against gas attack, filled the afternoon.

Day 9 was given over to lectures on the use of optically sighted rifles, including practising 'zeroing' the sights and the values of long-distance observation.

On **Day 10**, the duties of scouts, observers and snipers in attack and defence were discussed in detail. Grouping at 100 yards with telescope-sighted rifles was undertaken on the shooting range, followed by scouting in open country, and practical demonstrations of using scouts and snipers securing not only open country but also woods.

A lecture on front-line observation and reporting occupied the morning of **Day 11**, firing telescope-sighted rifles at 200 yards. Snap-shooting was practised at 100–200 yards with three-second target exposure. The afternoon allowed a directional march, with trainees put into four platoons and given map co-ordinates at which they must concentrate at a given time. Gas alarms were to be given and box respirators worn when necessary.

On **Day 12**, the duties of a battalion intelligence officer were explained in detail. Then came shooting at 300–400 yards and a period of German trench observation. The afternoon brought a demonstration of the use of scouts and snipers as a protective screen for the infantry in open warfare.

Understanding aerial photographs, assisted by lantern slides, was the key feature of the **Day 13**'s morning, including, where appropriate, study of the actual ground depicted on the photographs. Examinations in long-distance and front-line observation filled the afternoon.

Day 14 was 'open range day', while **Days 15 and 16** were devoted to scrutiny of students' notebooks, oral and written examinations, and shooting proficiency competitions.

Day 17, the last day, brought the course to an end, with the award of certificates to those who had made the grade.

APPENDIX 2

ROZA'S DIARY

Though not among the highest scorers, Roza Egorovna Shanina ('Roza Georgievna Shanina' on some documents owing to confusion about her father's forename) remains one of the best known of all Russian female snipers.

Born on 3 April 1924 on a collective farm in the Bogdanovskoy commune in the Arkhangelsk oblast, elder daughter of Egor Mikhailovich Shanin and Anna Aleksandrovna Shanina, Roza and her family moved to Edma when the commune collapsed *c.* 1933. She seems to have had a difficult early life. After undergoing basic education, but denied the chance she craved to study literature at secondary school, Roza left home at 14 – walking more than 200km westwards through trackless taiga forests to Konosha, on the railway line from Yaroslavl to Archangelsk.

Precisely why she undertook such a feat of endurance is contested. Her free-spirited approach to life may have been the catalyst, but there is a suspicion that her father may have blocked her progress: unlike her mothers and brothers, he merits not a single mention in Roza's diary.

Once in Arkhangelsk, she was able to attend secondary school and then, from 11 September 1941, worked at a kindergarten. War had begun by this time, and her eldest brother Mikhail was killed in December 1941 during the battle for Leningrad. Roza, a member of Komsomol since 1938, duly enlisted on 22 June 1943 after completing Vsevobuch marksmanship training, and graduated from the Central Women's Sniper School in the Spring of 1944 to serve on the Third Belorussian Front with the 707th Rifle Regiment of the 1138th Rifle Division of Fifth Army.

The snipers were noticed almost as soon as they had reached the front, a report that they had killed eight Germans being published as early as 7 April 1944. Roza was almost immediately exploited, no doubt at least partly because of her looks and self-proclaimed desire to be at the heart of battle, and her picture graced periodicals such as *Unichtozhim vraga* ('We shall destroy the enemy').

Aleksandr Stanovov, responsible for some of the earliest photographs, recollected that Roza initially refused to comply unless her fellow snipers were included, but the Soviet propaganda machine soon latched on to her, the writer and propagandist Ilya Erenburg doing much to raise her profile. Tall, athletic, blue-eyed and with cascading light-brown hair (sometimes described as blonde), Roza presented an ideal image.

Such celebrity inevitably caused resentment among other snipers in her unit, and even some of her friends were less than impressed. According to V. Mamonov and N. Poroshina, *Ona zaveshchala nam pesni i rosy* ('She Bequeathed to Us Songs and Dew', 2001), Kaleriya Petrova recalled that Roza 'walked a bit like a sailor, rolling here shoulders from side to side. She was well able to stand up for herself, she would be silent … but if you then went too far she would square up … and tell you precisely what she thought of you!' Others bemoaned her unpredictability, or belittled her rural upbringing.

Precisely how Roza came to prominence is uncertain, though she had struck a friendship with journalist Pyotr Molchanov, editor of *Unichtozhim vraga*, which may even have pre-dated her arrival at the front. It has been suggested that there was more to the relationship than mere acquaintanceship, and, if a list contained in the diary is believable, many men were involved in Roza's life. But the levels of this involvement – platonic or intimate – have been disputed.

Many have been quick to condemn Roza as a 'front-line whore', and it is fair to say that internet coverage is often prurient. However, there is no evidence in the diary to suggest that she was easy with her favours; indeed, certain passages can be read to suggest the opposite.

Roza Shanina is also often seen as an inheritor of the 'warrior woman' traditions of the Amazons, Nike (Goddess of Victory) and the many individuals who entered combat disguised as men. The journal *Rabotnitsa* ('Woman worker') depicted her as the female warrior of all ages: ethnic clothing and ancient armour paired with binoculars and a submachine gun.

Roza herself had no qualms about the sniper's role. She is said to have had 17 kills by 5 May 1944, her performance being recognised on 18 June with the Order of Glory third class and then on 22 September 1944 with the second-class award.

A total of 59 kills was entered on Roza's *Nagradnoy List* (official award record) late in December 1944. However, many more could have been registered by the time she was fatally wounded by shrapnel, dying on 28 January 1945. A 'final score' of 75 or 76 is plausible.

Roza Shanina was part of what she christened the *Ubegayutsaya troyka*, 'Runaway Troika', with Aleksandra 'Sasha' Yekimova (killed in February 1945) and Kaleriya 'Kalya' Petrova. The Russian word *troyka*, now commonly associated with a sleigh, can also mean a threesome. Serafima 'Sima' Anashkina, Evdokiya 'Dusya' Krasnoborova and Lidiya 'Lida' Vdovina, said Roza, also formed a Troika.

Memories Fade ...

Though Roza Shanina and others such as Lyudmila Pavlichenko enjoyed wartime fame, their prominence soon faded. Owing to a widespread perception of female combatants as 'front-line whores', many were destined to die in obscurity.

The acronym coined to describe them, PPZh, *Pokhodno Polevye Zheny*, 'marching field [or campaign] wives', a play on the designation of the Shpagin or PPSh submachine gun, had become common currency by 1945; and Marshal Vasily Ivanovich Chuikov could still write in *Nachalo puti*, ('Beginning of the Way', published in 1959), that the mobilisation of women on an unprecedented scale 'is still weakly covered in military literature, and at times unjustly forgotten ... Equally with men, they bore all the burdens of combat life and, together with us men, they went all the way to Berlin.'

Publication of Roza's diary in Russian in 1965, albeit in abbreviated form, inspired not only her own rehabilitation but also the rise of interest in other Soviet snipers at a time when Khruschchev's successors made Victory Day, 9 May, a public holiday in Russia. However, the view that women should not have killed men – even in circumstances as extreme as the USSR had faced during the Great Patriotic War – held sway for some time.

Though many reunions had been undertaken, Yuliya Zhukova recalling in her memoirs that 350–400 graduates of the women's sniper school had attended one such event held in 1975 in Moscow's Red Square, only when concerted efforts were made in the 1980s by Svetlana Aleksandrovna Alekseyevich and others to record female veterans' reminiscences did the situation change appreciably.

Lyudmila Pavlichenko's memoirs, incomplete on her death in 1974, were published in Russian in 2015 as *Ya – Snayper: v boyakh za Sevastopol i Odessu* ('I [am the] – Sniper: In Battles for Sevastopol and Odessa') and in 2018 in English as *Lady Death: The Memoirs of Stalin's Sniper.* Yuliya Zhukova's fascinating recollections appeared in English in 2019 as *Girl with a Sniper Rifle.*

Honouring her Words

Three notebooks that Roza kept from September 1944 to January 1945 – an earlier diary is said to have been lost – reveal a complex personality, balancing fluctuating feelings of inadequacy, and attention sought by all-but-reckless and sometimes insubordinate performance in battle, with the sensitivity that could manifest itself in poetry and song.

Roza's diary is presented here in a fresh translation taken directly from the original transcript of her notebooks. Undoubtedly a more accurate replication than previous versions, at least one of which contains errors suggesting use of an auto-translator, it gives a unique perspective of the war on the Eastern Front – far removed from the memoirs that have been re-shaped by the passing of the years, the need to follow the 'party line', and the fallibility of memory.

Roza's words are unfettered, echoes of a desperate time. Written in haste and never refined, her notes are not always coherent, reflecting moods swinging constantly from elation to despair and back again.

Adrenaline junkie she may have been, but in her words lie an underlying sadness and a desire to love that was never to be fulfilled.

And let them smile, secretly,
On reading my sad story ...
If I can't have love, if I can't have peace,
Grant me a bitter glory.

Anna Akhmatova 1889–1966,
from the poem 'Always So Many Pleas From a Lover!'

ROZA'S DIARY

October 6th, 1944

I met Gudkov, who was with Sergei [Shanin, Roza's brother] in the forests of Byelorussia; he asked me to remind Sergei of a little place called Kosino, where he was caught by a bombing raid while boozing. I am now in the editorial office. It's kind of strange when they have such stupid ideas about the forward area (Olga), even though they are only 25 km from the front. Yes, it's tough living in such surroundings! (By the River Šešupė).

October 8th, 1944

With Gudkov I went up for the first time in my life in an aeroplane. We are now in the 215th Rifle Division under [Major-General Andrannik] Kazarian. I pine for the 338th. The people there are quite different. I cannot do work badly; it destroys my conscience, but it's good, this messing around with the army newspaper. Out of envy the girls have spread rumours all around; my reputation is destroyed.

October 10th, 1944

The Germans have retreated; we are moving forward on the left. I met Kazarian. Everyone says he's good to the girls. He can be as unkind as he likes, so long as he's not a total 'womaniser'. I remember Gorodovikov. There's no comparison. How hard it all is …

Come what may, but I am not going to be the simpleton that everyone expects. I dreamt about my brother Fedya. It weighs heavy on my heart – I am 20 years old and without a good friend. Why? When there are plenty of lads around, but my heart doesn't trust anyone.

They say girls won't be allowed into Germany, and here we are already at the frontier. Where will fate land us? I remember Misha Panarin. What a good lad. Killed ... He loved me, I know, and I him. A senior sergeant, two years at an institute, a well brought up, simple, decent, nice guy. I felt very sorry for him. I can visualise Blokhin and Solomatin. I liked them, but I knew that our relationships were only temporary; they left and stopped writing to me – that's the proof. After the 338th Rifle Division I was with the top brass. In the 184th I didn't know anyone – just a little, superficially and on an ordinary comrades level.

How much injustice there is! Let's take the girls. A. Y. [Sasha Yekimova], my friend and sniper partner. When I'm successful, she's my friend, but when I'm not, there's no trace of her. At the moment I enjoy great authority, and she's with me. How I dislike this! I want friends like Agnia, who was in Grades 5–7 with me at school and Valya Chernyayeva, with me during the first three years at technical college. Such are not to be found.

October 12th 1944

I'm on the move, same as in July. We are heading for Sberki, 20 km to the left beyond Sloboda and the River Šešupė. We got into a vehicle without permission. It broke down near the 184th Rifle Division, and everyone went off to see friends, as it was close to evening. We spent the night in the dug-out belonging to the head of the political section. We were with Kalya Petrova, and ate the generals' dinner. They wanted to charge us for it, but we were not to be taken in. In the morning we were picked up by a jeep. We are on the move, but where to? It's a higher-echelon vehicle so the whole army is on the move, we established all that exactly. The girls spent the night on the front line, with troops advancing and firing, but at least they got to see the lads. I so want to be in the forward area – interesting and dangerous at the same time, but for some reason it doesn't scare me. I remember the days when I went into attacks with Solomatin, whom I loved, but did not believe in his love. He did everything for me. But death was looking him in the face. All the same I was someone to look after, and he was able to do this, maybe only because I was a girl and fighting courageously. I had only just left him when the outstanding regimental commander was killed right beside him. Nikolai S[olomatin] has now therefore taken over in charge of the regiment.

I left for the forward line. I met some lads, friends of our girls Shura and Dusia [Kekesheva], along with the battalion commander and deputy. I was splendidly received. I ended up in a company under a nice old fellow, a senior lieutenant and company commander. He took me under his wing. I went into an attack with him, running through some rye. Blokhin appeared out of nowhere. I found out that they had an attack planned that night and went off to join him.

At 3 o'clock in the morning they advanced. Firing all around us, and me in the front ranks. On seeing this, Blokhin turned to me and told me to go back. The deputy political commissar, the Jew Shapiro, sent me packing. It was getting light. I kept going. I was frozen. Where were our own troops? Fritzes on three sides. I took a look: there was a sentry in the distance, but whose? I crawled up through the rye. Our soldiers, an armed patrol, were tired and asleep in their foxholes. I ran up to the sentry. He was asleep on his feet. I learned that it was Solomatin's battalion. I lay down under some waterproof capes, facing the lads. In the morning they woke up and were surprised that I had found them. We're just sitting here.

Suddenly, a German plane started blazing away at the ground 100 metres away from us. Tairov said: 'In about 10 minutes there'll be an enemy counter-attack.' So it proved. We were ordered to occupy a knoll, I did so, and was at the very front. At first, I didn't see anything, and then I saw them: self-propelled guns creeping out from beyond the hill about 100 metres away with an assault force. I fired at the enemy troops. Beside me, to the left, a handful of metres away, a senior lieutenant and a captain and some soldiers were crushed. My rifle jammed. I sat down, cleared the obstruction, and fired again.

A tank was heading straight for me, about ten metres in front. I groped for my hand-grenades, then lost them as I crawled. I felt no fear. I'll crawl away, I thought. Virtually beside me our 76-mm gun was blown up. The tanks went past. Grenades were thrown from them and there was firing of every kind (machine gun, sub-machine gun, shells). Eight of them were knocked out. The rest turned tail. When it was all over and I saw the dead and wounded, it was terrible. The captain gave me his watch before he died.

We kept some booty as emergency supplies. For a long time, I held on to a blue silk scarf as a memento, then lost it. Tairov said: 'When the attack started, I remembered where you were, but you were lying somewhere up ahead and I was very concerned.' Tairov and Solomatin had an argument. Tairov is a veteran soldier, and gave orders to hold on to the last, or else they would surround us by morning, while Solomatin said: 'I'm in charge here.' They went off. I looked and saw General Babayan. I hid, so he wouldn't send me back to the rear. All the local Lithuanians were placed under guard. It turned out that an old woman was about to get a horse from the meadow when it was seized and sent to our rear. But when they captured the village where the Fritzes were, they found the horse there again. At night Solomatin and I remained encircled together. He is young … I wasn't scared of dying but all the same I burst into tears.

Fortunately, within two days, another division rescued us. I took my rifle and hand-grenades and went off 'to seek where in the world is a corner for strained emotions' [Aleksandr Griboyedov, *Woe from Wit*', act IV, scene 14]. There were Germans all around, both left and right. The gunner lads asked where I was going. I told them. 'Come with us instead,' they said, and I went. It was good being with them. We had long marches ahead and I got a lift on a gun. I had a letter from Blokhin to say: I'm in charge now, so go. We had a 60 km march. I got tired; we had hills to climb. I lay down thinking: the gunner lads will drop off to sleep and I will run away even though they're good guys and it's embarrassing to leave like this. They fell asleep but I was so exhausted I couldn't hold out. I was awoken by prodding. In front of me were two sub-machine gunners from the cadet company. I ended up with the 1136th Regiment in another encirclement. I spent the night there and, in the morning, I went to take a look. I noticed 30 Fritzes, who then ran off to catch up with the scouts. There was a tussle. Our captain was killed by two Germans with their rifle butts, leaping out of the bushes. He was about half a dozen paces from us, but the bushes were thick. We caught those two and shot them.

The Germans split into two groups and fled in two different directions. The lads ran after them, but I had to go 'home' to the company. I picked up a wounded man on the way. He asked me to shoot him there on the spot. I kept going. And in my dreams, I forgot that I was in a dangerous location. While crossing a bridge I casually gazed down at the overgrown gully below. I saw a Fritz standing there. It only took a casual '*Hände hoch!*' and six hands went up; there were three of them. One of them was gabbling something. I didn't understand; I only knew the words for 'Quick march' and I shouted them out. They crawled out of the gully. I took away their weapons, watches, boot polish, mirrors and so on. I escorted them about a kilometre and a half, looked and saw that one Fritz had only one boot on. He was the one who was talking to me in the gully. He wanted me to let him put his boot on. I had not understood. I met a lad – a soldier – who asked if I had a watch. I said: 'Here.' 'Show me'. 'Take it.' And he ran off with the watch. I escorted them up to a village, but the Fritzes had become quite bold. When, in reply to their question '*Gut* or *kaput*?' I answered: 'For you it will be *gut*,' they turned around and gave me a look.

I went through the village. We were in Poland now. There I was in my rainwear, with a Finnish knife, with hand-grenades, my rifle at the ready – looking like a bandit. The women stared at first but then everyone invited us for dinner. So much support!

I met Sasha Shchekochikhin there, and liked him. First, we went with Kalya Petrova to Blokhin's to have dinner, a drink of milk, and so on, but later I felt affectionate towards him and was embarrassed to say goodbye. We used to invite Blokhin to Sasha's, one way or another, but Kalya and I both liked Sasha. Realising this, Blokhin replied: 'He's busy,' although Sasha was free and glad to see us. I was the first to declare my love to Sasha in a letter, and therefore I was unable to reply to his positive response in a reciprocal fashion – I was ashamed. How I wept! When I left, when I captured the three Fritzes, and because I thought that he did not love me. I grew accustomed to him, thinking: this is the last time; the Germans will kill him, because this is a serious situation.

Now I can't love anyone. Although I don't trust Solomatin, I dream of meeting him; he is at hand. Blokhin is back in the rear, and calling me back there. I'm sure he's got Tanya there, whose letters I was not allowed to read. I correspond with Grisha, Dima, Kostya and Nikolai, but they are quite alien. It just happened, firstly out of comradeship and now the lads are dreaming of something more; they're bored at the front, and I don't want to offend them. Dima writes one, brief note in response to three of mine. I've asked him to send all my photos back. We both have feelings for them. There is nothing more to write really; everything of interest we have written down. How can I explain that I am quickly becoming disappointed in the lads? They are deceitful so at times I lose control and tell them to lay off.

I want to have a woman friend. I often think about Anya Smirnova and Masha Tisanova; I like them but I still don't know. They're not in our division. How can I explain that amidst such a multitude of men I am all alone? I don't know. If I had a guy, there would be separations and all sorts of unpleasant things. K., a fellow from the 215th Rifle Division, made me an offer. With perfume and anything I might like, but I am not for sale. I could take him for a ride, but it's not worth it. There would be unpleasant consequences; he's high-ranking.

October 17th, 1944

Action. We spent the night at Vovik Emelyanov's, with Sasha [Yekimova] and Kalya, but we also accidentally fell behind, as we discovered. There is a gap in the German frontier near the town of Naumiestis, to the left.

The tank crew invited us to join them and showed us how a tank is structured. What nice, sensitive lads! They all know me from the newspapers. I met the gunner lads who were there when five of our girls were killed all together by the River Neman. They saw that our lot was not easy either.

I am again to make one more break for the forward positions. I am even in tears at not being allowed. I want to go; how can I explain it? Some force draws me there; it's boring here. Some say that I want to be with the guys, but I don't know anyone there. I want to see real war. It's awkward, being a deputy platoon commander, or else I would have run off long ago.

October 18th, 1944

We looked for the Katyusha crew, but failed to find them. We spent the night in a different artillery battery. 'Attack!' We broke through the frontier. There we met Vanyusha from the 3383rd Regiment. What a meeting that was! We got carried away again. We found our unit, already making its way across German territory. There are prisoners, dead, wounded. We attacked a pillbox and took 27 prisoners, including 14 officers; they offered strong resistance. I am going 'home' to my unit. I can see the divisional headquarters. I got closer to the forward line and spent the night at [Sergei] Osmak's. I like him but he's very proud; maybe that's why I like him?

I met with General Kazarian and the head of the political section. I was genuinely in tears when they wouldn't let me go to the front lines. How can I explain that? On arriving 'home' I got a letter from Agnia Butorina. I always had good memories of this friend from primary school. She writes that her life is a mess and she's bored. I believe her: there are no guys around and it's no life for a girl. That's what it will be after the war. It looks like we are being sent to the rear. I dream of making a dash for the forward line.

October 20th, 1944

In the evening I slipped away to the forward line. I joined in an attack, and now we're stuck, consolidating. Rain, mud, cold. Long nights. But we are on the offensive.

October 24th, 1944

No suitable conditions for writing. I have been fighting. I went along with all the others; there were wounded and dead. I came because the regimental commander asked me to. Oh Lord, so many rumours! I remember weeping in the battalion, offended that they had seen fit to tell a rude joke in my presence. I regarded it as disrespectful. I recall my comrades who perished over this period. The same fate could lie in store for me, and I've reason to be grateful. Even my women friends have adopted an ironic attitude. The world is full of lies. It seems I don't have the strength of will to face this deceitful world for a whole lifetime. I got eight letters from Yasha Gudkov. Out of respect for his feelings I responded with one short letter, because he does everything he can for me. I am awaiting my photos; when I receive them, I won't write. Yasha has a true understanding of girls in the forces.

October 25th, 1944

Everything is good when you have at least some kind of woman friend. Sasha, I enjoy being with you from time to time. I share everything. Colonel Novozhilov summoned me with regard to the letter in which I asked to be sent to the front line and criticised our officers.

October 28th, 1944

A battle at Pilkallen [now Dobrovolsk, Kaliningrad Region]. Our troops have taken the town, killed the lot. There was a punishment battalion In the battle only one soldier has returned alive and unharmed. The rest perished.

I fought myself for a small patch near Pilkallen. So many times we captured it and were driven out! I fended off one counter-attack with great success. I wiped out 15 Nazis exactly, because I was at close range and did a lot of firing. Four gunners were observing through 6× and 10× magnification binoculars. When the Fritzes were crawling, only their helmets were visible and I fired at them. The bullets ricocheted. As they were tracers you could easily see them flying up off helmets. First at a range of 200 metres, and then closer, and at 100 metres they stood up to their full height, and when they were 20 metres from us, we ran for it. We lay behind a mound at the edge of the forest and easily concealed ourselves. Then we withdrew to a house. But the 'Slavs' all fled, leaving us on our own. Nearby Captain Aseyev, our battalion commander, a gunner and Hero of the Soviet Union, was killed. And we were the last to withdraw.

The order came to restore the situation. We crept up and occupied the house again, drove the Fritzes out. Then I went to the regimental command point, tired, and ate for the first time in a while. It was close to evening and I fell into a sound sleep. Suddenly there was firing right up close in the cellar. Fritzes, 15 of them, had crawled up. They were smashed by the gunners, who heard them from where they were, in a barn beside the house. The girls all turned out to be cowards, ran away. Only Kalya showed courage. Seeing the danger, the girls were ready to tear me to pieces because I was leading them to the front line. Sasha Koreneva was killed and two others were wounded: Valya

Lazarenko and Anya Kuznetsova. I am afraid to go home; the girls are putting all the blame on me. The soldiers, the lads and the officers, are happy with the courage I have shown. The news got to the top brass and I was recommended for the Order of Glory (1st Class) for repelling these attacks.

'In October 1944 the division received the command to mount a rapid offensive and take Pilkallen. The fighting was fierce. Despite heavy losses the enemy thrashed around like a trapped beast and continued to resist.

Our forward sub-units came to a railway line to the south of a settlement. In a small grove, by a forest warden's house, a small group of Soviet soldiers entered into an uneven battle with a similar enemy force. Also in this group was Hero of the Soviet Union Igor Petrovich Aseyev. (Note: war veteran Captain (Ret.) Medvedyev, who now lives in the city of Petropavlovsk-Kamchatsky, has told the story of this battle and Hero of the Soviet Union Igor Petrovich Aseyev.)

Towards evening the infantry and gunners, weary from battle and sleepless nights, came together for some time off. Volleys of six-barrelled mortars were unexpectedly heard from the direction of Pilkallen; around 300 sub-machine gunners were attacking the grove. The forces were uneven; the enemy outnumbered our handful of troops dozens of times over, and they began slowly to withdraw, to avoid encirclement. Captain Aseyev covered the group, maintaining continuous fire from a window in the house. Along with him was also sniper Roza Shanina. It was Aseyev's last battle. He covered the group's retreat at the cost of his own life.

There soon followed a command to advance. The troops went into the attack and drove the Nazis out. Our soldiers saw the prostrate body of Captain Aseyev.'

One eyewitness to Aseyev's last battle was Chevalier of the Order of Glory sniper Roza Shanina, who performed dozens of feats in the liberation of Prussia. She lost her life on January 28th, 1945. All that remains is a diary in which the battle at Pilkallen is very briefly described. Not everything is known about Aseyev and Shanina, and therefore it would be desirable if local history enthusiasts from the village of Dobrovolsk were to pursue further research.

November 5th, 1944

I haven't written anything for a long while. There was no time. I have been in action. I went with the girls and Vovik. They were gossiping that Captain Aseyev had been killed because of me, when I nearly perished on his account. I took a strong liking to Nikolai Ilchenko, lieutenant, gunner and brother of Hero of the Soviet Union pilot Ilchenko. He is in love with me. He pays no regard to my height, while I don't like it if men are even a little bit shorter than me, but, as it is, I suffer on his account.

I spent the night at Nikolai Fyodorov's. He's a good lad. He fusses over me; what I want I get (uniforms, caps, gifts). The only thing is I don't love him, enigmatic soul that I am, I'm fooling him, accepting gifts when I don't love him. Chekhov's 'Enigmatic Nature' [a short story, published in 1883].

The November celebrations are almost upon us. Invitations ... A timetable has been compiled, starting on the 5th ... But alas, November 6th–7th are working days for us, and some things have fallen through. There's a party on the 6th, the Katyusha crew. The tank crews have come, and Vovka Klokov. I wanted to go and see Borovik during the holiday period, but it covers working days. The Katyusha crew are all good guys. Only I love Vovka Letisov like a little boy, a little brother, but he's hinting at something; I don't like it. He gave me a Finnish knife. I don't understand anything at all, even life; everything's become so complicated.

November 7th, 1944

On the morning of the 7th I was in the forward area. In the evening of the 6th I had been to see Nikolai Fyodorov. It was fun, but there was an emergency. People were coming from Moscow. A press photographer was on the way, and the generals tried to summon me as a representative of the girls – a front-line sniper. But Nikolai didn't want me to leave and said I wasn't there. On the morning of the 7th I had a meeting with the generals and they dressed me down for not coming. I said: 'I wasn't told.'

Back home. More invitations, guys and declarations of 'my darling', 'my dearest'. To hell with it. I decided not to go anywhere, because I'm dirty and tired. Suddenly there was an invitation from the army HQ, from Molchanov. I couldn't refuse. They are good comrades and might think I've gone somewhere else. I had a cough but I went. I arrived, was sick both evenings, and I spent the 7th in bed. So, on the 7th I had half a day in Germany and half a day in Lithuania or the USSR.

I arrived home to find a pile of letters. But not one that I was looking forward to getting.

November 11th, 1944

I was summoned by General Kazarian. He told me off for poor discipline, leaving my post, going AWOL. But during the celebrations there wasn't a single problem. That's true even though nothing was put on for us in the division, and we had to celebrate somehow. What's to be done if the 'nobs' don't cooperate with us? Our section is operating normally. When I came home there was mud outside and Nikolai F. was sitting there. He had come to apologise for his bad behaviour, but there were a lot of people so he couldn't speak up, and went on his way dejected.

I have received letters from pre-school groups in Moscow and Arkhangelsk. They are all proud of my feats and my picture in the magazine *Yumor* ['Humour'] on November 7th. But they've exaggerated my achievements. I am only doing what every Soviet soldier is obliged to do, and no more. I am glorified everywhere but it's over the top.

November 14th, 1944

I acknowledge that I have earned fame in the army, well, at the front, but there's no point in broadcasting it across the Soviet Union, since I haven't done that much, have I? It's all talk. I'm going to bed. Many of the lads are burning with desire to see Roza … How can that be explained? I am depicted as a beauty or as a hero. Oh, and there was an incident today. At night there was a booze-up in the 277th Rifle Division. I met a Captain Lyosha, a nice guy, but he behaved like an idiot. Who does he think we are? I was invited by a lad from Arkhangelsk, a quartermaster. We gave them what for …

We've moved over to the 618th Regiment. Not well accommodated. The vehicle is being fixed up. Kalya from the 711th's quartermaster platoon and I are going to see Nikolai, who is in charge of this area. In the evening there was a talk on Stalin's speech. Didn't manage to see Trunichev, and he was so insistent with his invitations … There was an invitation from the commander of the 711th and, contrary to expectation, I ended up with the colonel in charge of the divisional staff.

November 18th, 1944

In a foul mood, but why? I have just been at Nikolai's, and spent the night there. My mood spoilt a little while I was still there. I met a boy in the artillery. And he was a boy. I took a real liking to him, a really sweet guy, and modest, but alas! Nikolai was beside me.

I recalled all my relations with Nikolai. I don't have much respect for him at all, but I still think he's decent enough. He appeals to all the girls. It's my bad luck that I became friends with him in such a mechanical way, not the way I wanted. I remember, the first day I met him during the offensive, when I 'ran' to the forward line. Tall, dirty, caked in mud and clay, in a long overcoat, like a real soldier. I respect him for his courage; he's a real Soviet soldier, but he doesn't stand out for his breeding and education – just an ordinary lad, a gunner. I recall the first days spent with Nikolai. Why didn't I have the courage to reject him? Conditions were cold and muddy; I had no warm clothing and needed help. He helped me; otherwise it would have been, in a word, ridiculous. I'm still a little bit fond of him even now, but for the rest I force myself, ram it into my head, to believe that I think a lot of him, and therefore I mope for long periods when I don't see him. Why do I emphasise the notion that I even love him? Because, after Blokhin, I have had no real liking for anyone and am incapable of it, but I don't want to be lonely; I want a friend, that's all.

I'm with the girls at the moment, bored. Tomorrow is Red Army Artillery Day. A secret commission has arrived to check on the behaviour of the girls.

Nikolai is suggesting formally that it would be easier to live together, to get married. What declarations of love Nikolai has been making! I still can't understand – is that the truth, or is he just a hypocrite?

Tomorrow is Artillery Day. Where will I end up for the evening? And will I end up anywhere? How hard it all is! O Lord! On the front line the situation is deplorable, and there's no more pleasure in it. The Fritzes roam far and wide, and shooting is not going well.

Dear Mum, I don't mind so long as you're not too depressed without me, and I find it depressing to have nothing to console me. How I want to see you!

I've got some news from Yasha. He was only with me for 10 minutes, and in that time they managed to spread rumours about me and Captain Aseyev. They won't leave him alone even in the after-life – poor guy. Yes, I received a certificate of honour from the Young Communist League. A letter arrived from a girl I don't know, who, like me, is burning with desire to get into the heat of battle. I told her everything she wanted to know.

November 20th, 1944

There were so many invitations to parties yesterday (the Katyusha crew, the commander of the 711th, Trunichev, the reconnaissance lads, the 120th Battery and many, many more), but I decided to turn them all down and go to Nikolai's for the evening, although I knew that I would be spending it in a more modest fashion than anyone else.

I am sitting with the artillery major. I like him as a person – he's modest and kind. But Nikolai S. is just a child. The major instantly fell insanely in love with me and the lieutenant in charge of artillery reconnaissance was there raving in his sleep. This is the first time I have played so unkindly on the nerves of these two Nikolais and some others, the first time I have been so hounded by declarations of love from the guys and duels on these grounds, and where did it happen? In the forward area.

November 23rd, 1944

Yesterday evening the Katyusha crew came. I freaked out a little, jumped out of the cab and ran off. I was in tears for a long time. I walked about 15 km, came to my senses when I got to the forward zone, then got confused because I couldn't see a damn thing. I felt my way along, crying. I don't like this life of tripping around. After all, we're at the front and we ought to operate as if we're at the front – with less attention to extraneous matters, but what are we doing instead?

I wept with all my heart the whole way. Because it was tough. I was alone at night, with bullets whistling past, fires burning. I came back, lay down to sleep and slept till two o'clock the next day.

Last night and today I have been writing letters. I got in touch with Nikolai F. I received something from Lukyanyenko and some of the other members of the tank crew. They all know me and remind me of my cheeky laughter with the song 'On Tramped the Fritzes, Sewing up their Britches' and look at my photo in *Krokodil* magazine. I haven't seen it. When they see my photographs in newspapers or magazines the lads either carry them in their map cases or hang them on the walls. There seem to be more who respect me than detest me.

November 24th, 1944

On learning about his departure, I went off to spend the night at Nikolai's, but not because I'm sorry to part with him, but because I need some things: a waterproof cape, a book, and a watch as well, but I didn't take the watch. The girls aren't there – there are ten of them in different divisions. We are going off to the 203rd Reserve Regiment. Pretty-boy Nikolai Borovik won't be meeting me any more – he's fighting somewhere south-west of the town of Naumiestis.

We've reached the divisional rear area and are spending the night with the cadet company. We've been quite well put up.

I met the general's suite and had supper, but I'm in a foul mood. There's an accordion playing, like being in quarters. Still, it's not really the front, and everything's quite different, everything according to the way things are done in the rear. How nice! In any case a girl is well received everywhere: 'Come to us ...' O Lord, what a bore, just somehow boring, boring. How can I explain it? Still, whatever the case, they're friends and it's more fun with them. It's pleasant all around. I felt like dashing over to see the lads, Nikolai Ilchenko or Borovik, but would sooner see Shevchenko and, well, call on Fyodorov, but no, not at the moment.

November 26th, 1944

Set ourselves up with the 203rd Reserves. It's not bad. We are relaxing.

November 27th, 1944

I got my first glimpse of German *Fraus*; didn't like the look of them.

Yesterday there was a dance. I'm not much of a dancer, but they inspired me – K. [Kalia?] is concerned about fashion. They wanted to go for a spin in a jalopy. Vovka came, but it didn't work out. They wouldn't let him through. Surveillance, and more out of envy. Coming back from the bath-house, we remembered how the Germans had dragged away our girls. Dusia Kekesheva saw the whole thing with her own eyes. She herself escaped their clutches. Shambarova pretended to be dead, and there are two more somewhere, possibly still alive. In the hands of butchers. We should now take it out on the German women, but I haven't the heart to. I feel cool about everything.

'Dear Roza,

On the night of November 30th, 1944, I had a dream:

"It was June 1947. It was a fine sunny day. The capital of the Soviet Union, Moscow, was enjoying its former bustle. Two years had passed since the end of the war. The war veterans had returned to their home patches. The doors of tertiary institutes, technical colleges, and institutes lay open before us. How much these people had changed in two years of peacetime life. Today it is especially noisy on the streets of Moscow – it's Sunday. Two girls are walking along Gorky Street. Each has a bundle in her hands. It's noticeable that they are hurrying somewhere. On reaching a white house, they stop by the entrance. A few minutes later a third girl emerges from the entrance. Her lips bear a glowing smile. After only a few seconds this trio continue their walk. A discussion breaks out among them. Their words are accompanied by gestures. None of them wants to concede a point to the others. Finally, the dispute comes to an end. Exhausted by the heat, they slow their pace. Silence descends. Coming to meet them is another couple. In an instant the eyes of the girls are fixed on this couple. The slim young man in military uniform is elegantly dressed. Wrapped in a light cover in his arms is a child. Beside him walks a slim brunette. They

are chatting away merrily about something. Looking at them, it is possible to draw the conclusion that they are happy. The young man directs his gaze at the three girls. Their eyes meet. In an instant the whole sequence of their lives at the front passes before each of them. Their friendship as snipers and Katyusha rocket crew.

"On drawing level, they do not stop. Only nod as a sign of greeting. Two years of peacetime life have radically changed these people. It is difficult to recognise them. The girls are studying at an institute, while the young man is employed at an academy and has a wife and son. It is their first meeting in the two years since the war ended. In two years, relations between them have changed. But at one time they were great friends. So, why do this young man and these three girls not stop on meeting, not recall the arduous days of war, when everyone shared the burdens and adversities of wartime? Now they are a long way away from that. Seemingly the young man considers it beneath him to admit to being their friend in front of his wife. The three girls continue on their way, but in silence. If you look them closely in the face, you would have no difficulty noticing that one of them is particularly gloomy. It would appear she found the encounter traumatic. Such is the way good old front-line life comes to an end."

Roza, if we stay alive and well, but end up in different parts of the country, I beg you: don't forget Kalya and me, and our valiant itinerant trio.

A. Yekimova.'

November 29th, 1944

While describing this dream, Sasha Yekimova read my diary. She spoilt both my mood and her own. Yes, she was now very dear to me as never before. Say what you like, but she shared the whole soldiering experience – both joys and woes. No person is free from sin; I do not condemn her for that negative attitude which I have noted. But the atmosphere! Frightful!

I am back on the newspaper, so at the moment I am maybe 7 km away from her. We wanted to leave, but can't, not till tomorrow. What music! The nicest things are broadcast by radio. I sincerely wrote to Agnia Butorina that I have no ambition to meet anyone because every minute brings fresh news.

All in all, my future is not certain; there are numerous possibilities: 1) enter an institute 2) if that first option doesn't come off, in which case I will become a public servant, devote myself wholly to rearing orphans; that's my special field. In the case of the second variant I shall also study something, to raise my general level in all respects. All in all, I have not mapped out my future, not had a good hard think about it – it just occurred to me.

I wanted to study communications, Morse code, and so on, while I was here, because there are courses for signallers next door, but we are leaving. I want to have many different special skills, although not to work in them, but just to know them, to be able to help out in an emergency. Well, enough dreaming, I'm fed up with it for today.

December 2nd, 1944

O Lord, how tedious things have become. I arrived at the general military stores base. In anticipation I had remembered absolutely everything. In particular, two pictures passed before my eyes

Pavel Blokhin lying in a dug-out in the 36th Regiment, 338th Division, near Vityebsk, with the telephone receiver in his hands roaring away: 'You mother …!' I am sitting beside him. He puts the receiver down and smiles at me.

Nikolai Solomatin and me running through the forest by the Neman, through the bushes down the slope of the bank, running quickly. I feel my head: there is no green camouflaging headscarf on it. It is a hot sunny day. I run on further. Nikolai looks around: it will be hard to get up the steep ravine. He takes me by the hand, helps me to scramble up and kisses me firmly, and we run on further. We come out onto the high, high bank of the Neman – with a field to the left, forest in the distance, the river to the right, and beyond the river – meadows and woods. We proceed as quiet as quiet, exchanging glances. Suddenly a machine gun starts rattling away to the left – it's a Fritz. We quickly jump down the bank into the bushes and carry on through them.

Another time, it was night. Nikolai and I are walking towards some village, in the direction of the Germans. We trudge by a forest all night, guided by a Lithuanian. We go round the whole forest, following a stream, and come out on a high hill. There is farmland there. We lie down to rest under the bushes on a waterproof cape, then get up, heat some coffee, boil some soup and have something to eat … beyond that I don't remember which direction I followed. I remember that beyond the village there was a fierce battle going, but more than that I don't recall.

I remember the big march. It was raining, I didn't even have an undershirt on and got soaked to the skin. He brought his cloth uniform to put on, but I didn't take it; I took only a cape. At night we got soaked, ended up in a puddle, such was the rain. He and I spent the night in a buggy. I like him so much! I sang the song: 'Where have you been all year, where, where, where has the war cast you …'

So boring! An accordion is playing in the workshop. What a drag! I want to go there now. Forward! To where the fiercest fighting is. I don't want anything else? Why can't I do it, eh? How oblivious these commanders are! I'm going to finish writing now.

December 3rd, 1944

I've lost Nikolai S's address. It's so upsetting I have automatically lost trace of the one whose image warmed my heart. His cap often reminds me of him. It's a good day. Oh, what a pain, what a pity to lose N. S.'s coordinates. There's some extraordinary music playing. My heart stops at all the recollections. I am tearing up the letter which I wrote to him and could not send, not knowing his address. My heart craves tenderness, guys who are familiar to me and respected by me. They're not even on the horizon.

I am sitting at Captain Sokol's. Again, I cherish the hope of getting into the 338th Rifle Division, although today I was dismissed by the battalion and company commanders because they have got sick of me bothering them with this request.

December 4th, 1944

Been to see the major, captain and battalion commander; with him was Major Lyapyenko. Sasha's and my relations with them are not all straightforward. We sat from 6 to 11 and missed the dancing. They arranged for us to stay the night, but we didn't stay. It would have been unseemly, and today we are regretting it – it was our last evening. They played good music but we waved one another goodbye without even shaking hands. A lot of the high-ranking personnel were there.

Today we did 20 km on foot – mud, wind and snow. We met the Katyusha crew, and went to their base for the night, although the platoon commander did not give permission. But there were no provisions there, nowhere to sleep. Sasha and Kalya went to Vovka's on night patrol, and I bedded down in their place. What modest lasses!

December 6th, 1944

Good God! Surely you won't refuse to help me to sort everything out? It's all so tangled, O Lord! I have written 30 letters to all destinations, both business and ordinary ones. I didn't sleep the whole night. Only in the evening, and then I woke up, had a total rethink of my life, about all the girls, about fairness.

You must know that in the course of my life at the front there has not been a minute when I did not thirst for action, wanted hot action, wanted to go along with the soldiers. I am sorry now I was not born a boy. No one would pay me any attention, no one would pity me, I would fight with all my heart. Now a question has arisen. I said: 'I want to take part in the next attack. Kalya and Yeva believed me; they know my nature. All the rest said: 'Stop fibbing.' But Yeva told the girls that she had heard from the soldiers, how I had personally lain down in front of German tanks, and that she completely believed me, for that was voluntary as well.

I would give everything to go with the soldiers on the offensive. O God, why have I got such an enigmatic nature? I just can't understand. I thirst, thirst for battle, the heat of battle. I would give everything, including my life, if only I could satisfy this whim. It torments me. I can't sleep peacefully.

I'll write about the platoon. Alkimova says: 'I don't believe Roza has wiped out so many Fritzes. They've just been attributed to her.'

It works out like this. On defence sometimes you do a lot of shooting at targets, but it's a mystery whether someone has been killed or not. It's reasonable to suppose that I always hit the targets exactly and also scored Fritzes more often than not when they were stationary; in most cases I fire at stationary figures and troops on foot. It's difficult with deserters: all you can do is scare them. Sometimes they don't record things at all, and sometimes they record on the off chance, sometimes without foundation, but in my tally, there is not a single falsely recorded dead Fritz. If one was recorded without foundation, then the next one I killed wouldn't be recorded as required.

I remember one time on the offensive, and I can say it with my hand on my heart, sincerely. I was helping to beat off counter-attacks. I expended about 70 rounds. In that attack I finished all of them off. One man took to his heels, but the others were immobilised and wounded just from soldiers' bullets. I was using an old Berdan rifle covered in mud, which hadn't been tested for a year and was not set up for targeted fire. But I was firing from distances of 50 or so metres, right up close, and I wounded another 20 at least. While on the offensive you often had to fire from close distances and be on the mark, and I didn't miss.

I remember the last counter-attack; the bullets were landing right on the helmets. If a helmet appeared, tracer bullets would not penetrate it beyond 100 metres. They were ricocheting upwards and it was clearly visible. It was then that the lads witnessed the accuracy of my firing. I was the only one shooting, and there were five of them observing and saying: 'Well done!' I hit a Fritz standing upright 20 metres away and clearly killed at least 15, or even more. So, in two successful hunts that amounted to 35 Fritzes. A sniper's aim is true.

But on defence I often operated beyond 50 metres and fired at the Fritzes full-body, and at the chest: 57 dead Fritzes and not one of them credited. They can say what they like, but I know; the doubts are all due to Sasha Emelyanova. There are no folk without negative traits. I don't blame her; she's just brought up that way. She's used to thinking of herself and nobody else. To a small extent she's become a different person. I've got used to her, become attached to her and Kalya. It's tedious without them. I respect them more than anyone in the platoon. Life's still easier with friends. The three of us are from quite different families, though we've all had one year at an institute, and we all have slightly different characters. Our life on the home front and backgrounds are different, but we sort of have something in common. We are friends, and firm friends at that.

Kalya Petrova is also my friend. She's a good girl. She is devoid of egoism and bold, very common-sensical, copes well with all issues. She has a fabulous memory, but is inclined to be a bit on the lazy side. Sasha, Kalya and me – we form an itinerant trio of friends.

All the girls are more or less pleasant; there are no beauties and no freaks. I like Sasha and Masha most of all; they are not beautiful, but appealing. Only Nyusska is sort of 'old world'. Lelya is about 22, Nyusska about 23, and the rest are 24 or 25. Such is our section. I've been writing all day. I'm tired. I'll finish this later. I have written all my letters and my diary, and all on my knees. Then I wrote leaning against the wall, and my back and my hand got tired.

December 7th, 1944

I went to the bath-house and checked the sighting of my rifle, as did the platoon, and that was it for the whole day. I wrote a letter to Comrade Stalin, asking for a transfer to a battalion. I want to take part in an offensive. O Lord, the Fritzes are beginning to play pranks. They snatched away a battalion commander from the 277th along with his deputies – rich fruit for interrogation. One regimental commander has ended up in a penal battalion. Snipers are good providers of information if taken; we know all the armies, not only the divisions, and can locate them in any area. Yet Nesterova and Tonailova did not say a word when the Fritzes tortured them – good on them, even though they were called collaborators. I saw their photos in a German newspaper, although they were from some time back, and taken from their Red Army booklets. I'm finishing up and going to bed.

Yes, I remember my beloved Arkhangelsk, the foreign seamen, the foreign sailors' club, the international club, the Dynamo stadium, the Grand Theatre, the Edison, Ars and Pobyeda cinemas – all the cultural and recreational centres. I can picture them now.

Zina Andreyeva is from Alma-Alta, Tosya Kot from Kuibyshev [now Samara], Anya K[uznetsova] from the Tatar Autonomous Republic; Tamara Alkh. and Kalya are from Moscow, from the Taganka and Kropotkin suburbs. The rest are from Sverdlovsk [now Yekaterinburg] and Molotovsk [now Perm] in the Urals, Arkhangelsk, and Siberia – Masha Rozhkova is from Omsk. That's all for now.

December 9th, 1944

Everything's not so bad. In the evening of the day before yesterday I went to listen to the gramophone at the major's. He's the head of the operational section. The records were sad: 'Hour by Hour', 'The Coachman', and others; I was overcome by a kind of melancholy. At first, I restrained myself, but then I couldn't any longer. He began to tease me. I burst into sobs, cried so much and played 'Hour by Hour' about ten times. I wrote to Comrade Stalin again. The lads from the neighbourhood write sensitive letters, make all sorts of declarations, invite me out, but I have decided not to go anywhere. I write to everyone: 'I can't,' sensitively explaining that I'm not like everyone else. They don't get any attention, and when I leave, the deputy platoon commander will say: 'She's a heroine and all that, better not to get involved.' I sit and mope, and I haven't seen the lads for a long, long time, even though we live close by – 3 km apart. I write shortish, but very frequent, messages home on snaps and postcards.

December 13th, 1944

There was a meeting of the women in the division in the evening, the day before yesterday. A lot was said about me, as a good example. Our snipers put on a concert, which wasn't bad, with Zoya Mikhailova as compère. On the 12th I was wounded. It's amazing; I dreamt that I was being wounded. Later I was sitting in position with my sights, I remembered the dream, and it seemed to me as if I had been wounded in the right shoulder. Not even five minutes later a Fritz sniper bullet hit me just where I had dreamt the wound. Moreover, I didn't feel any particular pain; my whole shoulder went numb. It was bandaged up and, without demanding special care, I went home by myself. I didn't want to go to the medical centre, but I was forced to. The operation was painful and I wanted to go home. I wasn't allowed back to my unit; they slashed me around. The wound, it seems, was trifling – just a couple of holes – but the way they cut me up, it won't heal for a month.

I am now in hospital. The joint is sore, the whole shoulder, but not that much. I am thinking of running away. What will happen next, I don't know. I took a liking to one nurse, and the doctor is a woman, a major, but here they're all ignoramuses. I got an invitation from the sniper lads to their party, but alas, I'm under treatment. I'm not home and they'll come for me.

December 18th, 1944

Every day I dream about my friends: Sasha and Kalya. How I pine for them! I'm getting a lot of letters; the girls bring them (from Karshinov, Borovik, Rumyantsev). They're all head over heels in love. It's only to Vanya that I write nice letters, from the heart. He's a good lad, a senior sergeant. I get them from the girls; we studied together. They congratulate me on my achievements.

I'm now just back from seeing *Lermontov* [poet and painter; presumably the 1943 biographical film] at the cinema. What a big impression it made on me! Lermontov's character is just like mine. Following his example, I have decided to behave not according to anybody's requirements, but the way I want to. His silhouette on the iron bridge, the horseman, will remain in my memory; I also want to be first in something. Now nobody will persuade me, and if things are bad for me, I too am capable of resolving the issue at any moment, for in no way is my life what I want; I simply exist. So, let me give them something to remember me by, so that not just some people but lots of people know. Oh, how Lermontov's character appeals to me!

December 19th, 1944

I'm in an army convalescent home. Yes, I do want something – not just to stand out, but to satisfy some need that torments me. I'm still in the same mood. I was driven here in a jeep; some major was heading here, so I was given a lift. The driver, it turned out, lived in Arkhangelsk, right close to me. He promised to give me a sniper badge which was awarded to him by mistake. He's from a tank crew. And again, there's a picture of me in *Frontovoy yumor* ['Front-line humour'] magazine of November 7th, 1944. O Lord, what a bore! We've been quite well accommodated. I'm now going to read [Theodor Dreiser's] *Sister Carrie*.

December 27th, 1944

When life is going well, you don't feel like writing.

I have been in a convalescent home, read *Sister Carrie* and *Bagration* [probably *General Bagration* by Sergei Golubov]. It's a good book. 'O Carrie, Carrie! O blind strivings of the human heart, onward, onward, it saith, and where beauty leads, there it follows.' You read and you wonder what Theodore Dreiser or Bagration have to do with you, what does glory mean – either splitting open your own skull in the name of the motherland, or smashing someone else's. That's real glory! That's what I will do, by God. I have seen a lot of films: *In Old Chicago* [20th Century Fox, 1938], *Zhdi menya* ['Wait for me', Mosfilm, 1943] etc. *Podvodnaya lodka T-9* ['Submarine T-9', Lenfilm, 1943] impressed me more and it made me think a bit about it, but the others were so-so; I wouldn't call them good.

At the moment I am sitting at Nikolai F's. I just dropped in and think it's for the last time, because it doesn't appeal to me. It's already a month since I called on him and, afterwards, I didn't even talk to anyone. The document is being prepared for the award of the 'Order of Glory (1st Class)'.

Yesterday I was pursued by a nice guy: 'Let me kiss you; I haven't kissed a girl for four years,' and he begged so insistently that I gave way, and he really was nice, not at all repulsive. It was pleasant, damn it all. 'Kiss me, but just once,' I said and was almost in tears myself. Why? Out of pity.

'Yet true love is no wanderer; it gives a beauty that it does not find, and forges a chain no charms can break.' (*The Heart of Princess Osra*) [by Anthony Hope, 1896]).

'O, passions, passions! O, blind strivings of the human heart. Onward, onward, it saith, and where beauty leads, there it follows' (adapted from *Sister Carrie*).

'Glory means either splitting open your own skull in the name of the motherland, or smashing someone else's …' (*Bagration*).

'Am I to keep my will within legal bounds? The law makes crawling snails of those who would soar like an eagle' ('The story of my life'. The words of the rebels [source unidentified]).

Beloved of mine

You write, offended, telling me that I've forgotten you,
But you forget that I'm at war, beloved friend of mine,
And more than I can count await a letter from me too.
In Omsk they wait, and Tomsk as well, beloved friend of mine,
And a lawful wedded wife as well who's waited many a day,
So, I'm destined to forget you, beloved friend of mine.
We have a child who looks like me, a daughter, so you say,

So, may she thrive, I do not mind, beloved friend of mine.
If asked who is the father of this child I never saw,
Tell them that he's at the front, beloved friend of mine.
Forgive me for this jest, my love, the culprit is the war,
And wait for me no longer now, beloved friend of mine.
Although I'm very proud of you, my family's first in line.
To you I shall no more return, beloved friend of mine.

[Author unidentified]

Midnight. The candle is guttering.
Stars can be seen through the door.
You're writing a letter, my dearest,
To the flaming inferno of war.
How long you've been writing it, dearest.
Completing, you'll start one anew.
Yet I'm sure that, to the front line,
Your love will come bursting through.
Far off, the lights of our quarters
Are hidden by wartime smoke.
But, whether at home or the front line,
We're loved and remembered by folk.
Letters bring warmth to the front line.
Reading them word by word,
You hear the voice of your homeland
As a voice through a thin wall is heard.
Soon we'll return, I believe it.
That glorious day will dawn.
Gone will be grief, separation,
And joy in the home reborn.
And then sometimes in the evenings
We'll cuddle up closely and then,
Retrace through our letters war's saga
And relive our feelings again.

[Lines by Joseph Utkin, 1903–44].

Greetings, my belovèd, just as ever,
I wish you no offence, so never fear.
I still remain affectionate and tender.
Only my expression's more severe.

It's no bad thing my character is stronger.
My language has become a little coarse.
The laws that war instructs us in are different,
And now I am accustomed to its laws.
Here we're daunted not by freezing weather,
Nor squalls of flaming fire, red and bright,
Yet I remain the man that you remember;
You'll recognise me easily on sight.
Amidst the sound of shells and bullets whistling
I march again today into the fray,
Clad in what was once a newish greatcoat,
The one we both recall from yesterday.
Whole-heartedly embittered at the enemy,
I'll prove myself a gallant soldier true,
So that once again a life of freedom
May be shared in joy by me and you.
Meanwhile, my beloved, just believe me,
The roar of battle's sounding far away.
It's time to go, to head where death is stalking,
Cross fields of fire and enter in the fray.
And while I am returning from this battle,
Wait for me, and do not mope in vain,
And I'll reward you with a different greeting
And greet you with a burning kiss again.

[Author unidentified]

If the heart is filled with sadness,
If a fearsome battle's due,
May you visualise, remember
All there was between us two.
Do not entertain the notion
That we won't meet again one day,
And all your happy memories
Take with you on your way.
And, leaving for fields of battle,
Through regions soaked with gore,
Remember, my love will follow,
Follow you ever more.

[Author unidentified]

January 8th, 1945

There hasn't been any paper so I haven't written for a long time. After the convalescent home I went to see General Ponomaryov, who is a member of the Military Council, to achieve my specific goal of getting sent to the forward line. I was sent to the commander-in-chief of the 5th Army, Colonel-General Krylov. With a lot of effort, we were able to arrange to see him on February 5th; he would give me a document allowing me to take part in the offensive. I managed to get some good uniforms for the girls; they are poorly dressed. Ponomaryov ordered everything to be done, but now I see that his command is not being carried out with any enthusiasm; it's all just words. I sat the whole day in reception for Ponomaryov and nevertheless gained an audience. I was lightly dressed, and I was given a sniper's fur coat and was specially taken by jeep to the medical centre. I'm happy. I only got fur coats, felt boots and camouflage dress. I froze on the journey, but at least I achieved that. It's cold in the trenches. My overcoat is now too short; I've grown over the year, and I don't like it anyway. I've swapped it for a wadded jacket.

I ended up with the girls in the 157th. I wouldn't recognise them! My friends Sasha and Tosya have got married. Lord, that leaves four girls, no, five, out of twenty-seven. And in my absence one good girl, Tanya Kareva has been killed. Her friend Valya L[azarenko] came back from hospital after being wounded, and what a surprise! No Tanya. Valya is weeping. Valya is twenty-five, a good girl, with character, pleasant, big and blonde, and mad about horses. She is the most independent girl out of the whole platoon. She completed seven years at school, worked at a factory and did a factory training course.

Well, so long, I'm going to get my watch fixed. When it gets a bit warmer, I'll run away to the forward line. For repelling counter-attacks (It was really torrid and I risked my life) I have received an award, the medal 'For Valour'.

January 13th, 1945

I didn't sleep all night, felt bad; I've fallen sick. The Germans pounded us solidly. Today from 9 o'clock till 11.30. Our own artillery preparations went on and on. [A major Soviet offensive in East Prussia began on January 13th.]

The Katyusha rockets gave the first signal. They made it hot for the Fritzes. Now the situation is still unknown. I've got travel fever. O Lord, the noise in the dug-out! Dug-out life goes on. Smoke. The bunks have been changed from double to single layer. No sooner have you finished work than you have to move on again. There's been a bit of frost; your feet are damp in felt boots and cold in ordinary ones. They are all taking their sheepskins. I come from up north and don't need one – makes it difficult to walk.

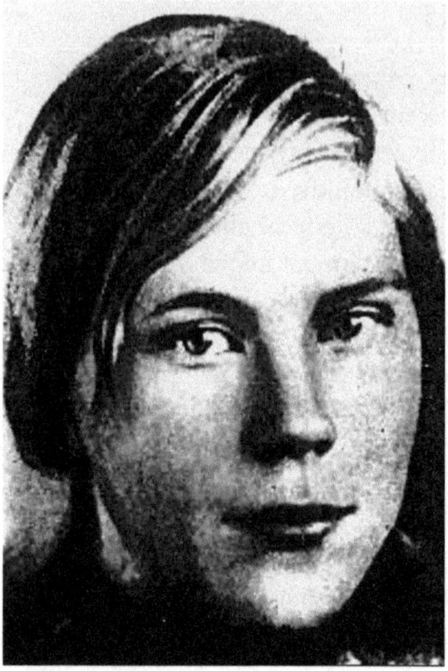

Roza Shanina in the summer of 1942, before her military career began.

A draft of women snipers at the station, ready to leave for the front at the end of their training.

Kaleriya Petrova recalled that Roza, shown with her M91/30 rifle and PU sight, 'was well able to stand up for herself; she would be silent … but if you then went too far she would square up … and tell you precisely what she thought of you!'

Roza Shanina wearing the Order of Glory Third Class awarded on 18 June 1944. The photo was taken soon after the award was authorised at the *sovkhoz* (state-owned farm) in Krynki in what had been Poland prior to 1939.

Roza with her friends Faina Yakimova (left) and Lidiya 'Lida' Vdovina (right).

Roza Shanina wearing the Orders of Glory, Second and Third Class. (*Arkhangelsk Museum Archives*)

The sniper's greatest fear was enemy mortar fire. In order to dislodge the Russian snipers, the Germans would rake the terrain with artillery fire.

Snipers Roza Shanina and Aleksandra 'Shura' Polygalova discuss their art with Major-General Basan Badminovich Gorodovikov, commander of the 184th Rifle Division.

(*From left*) Roza and her comrades Lida Vdovina and Sasha Yekimova.

Roza Shanina and the officer commanding her unit, usually identified as Major Aleksey Balayev, practise with her M91/30 sniper rifle.

Since she was both a successful sniper and an attractive and photogenic young woman, there were many photos of Roza taken for publicity. This one was shot in the village of Stolerichiki in 1944.

Roza Shanina wearing her Excellent Sniper Badge and the Orders of Glory, Second and Third Class.

Taken from the propaganda newspaper *Frontovaya Illyustratsiya* ('Frontline Illustrated'), this group photograph of Roza's 'sniper commando' pictures, from the top left, Mariya 'Masha' Rozhkova, Roza Shanina, Olga Mokshina, Yeva Novikova; Antonina Kuznetsova, Aleksandra 'Sasha' Yekimova, Evdokiya 'Dusya' Krasnoborova, Antonina Pryalkova and Zoya Shmeleva. Credited to Belyanin, it was taken on 20 January 1945.

Senior Sergeant Roza Shanina poses as though preparing to shoot while Zoya Shmeleva acts as observer. Third Byelorussian Front, 20 January 1945. (*Belyanin photograph*)

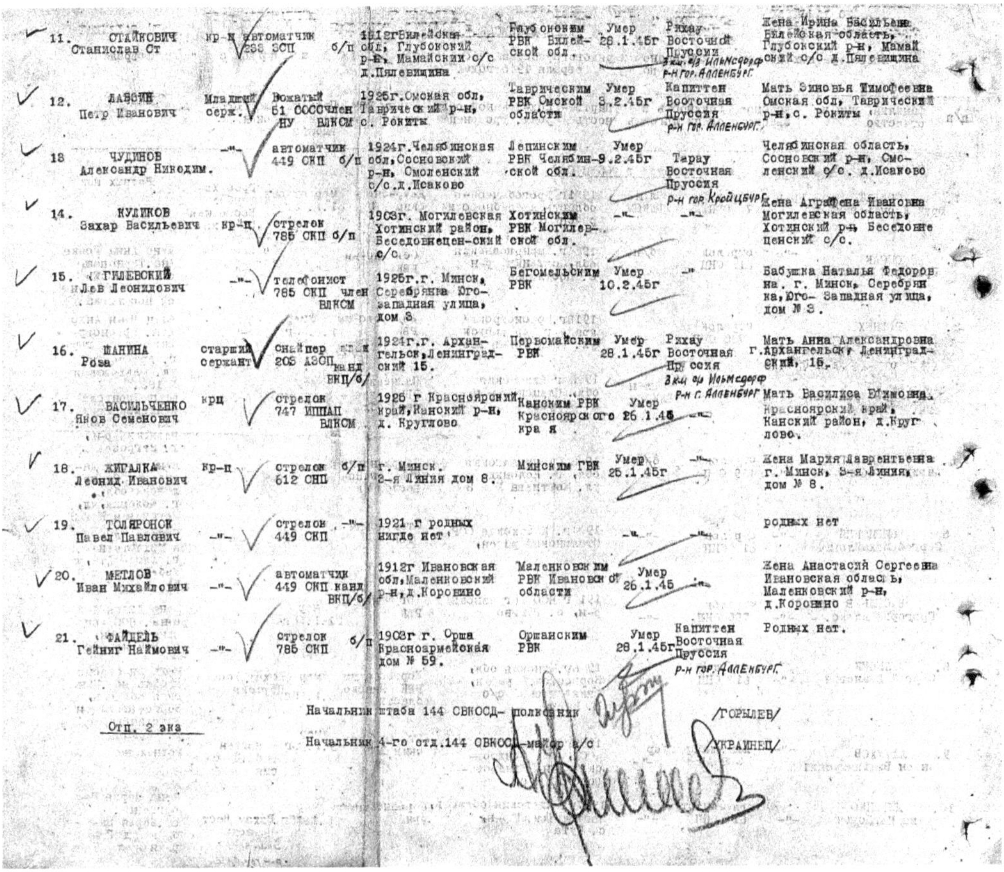

No.	Name	Rank	Position / unit	Born / address	Called up by	Fate	Buried	Next of kin
11.	СТАЙКОВИЧ Станислав Ст	кр-ц	автоматчик 238 ЗСП б/п	1912г Вилейская обл, Глубокский р-н, Мамайский с/с д.Пялевищина	Глубокским РВК Вилейской обл.	Умер 28.1.45г	Рхау Восточной Пруссии 3км. с/з Ильмсдорф р-н гор. Алленбург	Жена Ирина Васильевна Вилейская область, Глубокский р-н, Мамайский с/с д.Пялевищина
12.	ЛАВСИН Петр Иванович	Младший серж.	Вожатый 61 СОССчлен НУ ВЛКСМ	1925г.Омская обл, Таврический р-н, с. Рокиты	Таврическим РВК Омской области	Умер 3.2.45г	Капиттен Восточная Пруссия р-н гор. Алленбург	Мать Зиновья Тимофеевна Омская обл, Таврический р-н, с. Рокиты
13	ЧУДИНОВ Александр Никодим.	-"-	автоматчик 449 СКП б/п	1924г.Челябинская обл, Сосновский р-н, Смоленский с/с.д.Исаково	Лепинским РВК Челябинской обл.	Умер 9.2.45г	Тарау Восточная Пруссия р-н гор. Кройцбург	Челябинская область, Сосновский р-н, Смоленский с/с. д.Исаково
14.	КУЛИКОВ Захар Васильевич	кр-ц	стрелок 785 СКП б/п	1903г. Могилевская Хотинский район, Беседовнецен-ский с/с	Хотинским РВК Могилевской обл.	-"-	-"-	Жена Аграфена Ивановна Могилевская область, Хотинский р-н Беседовнеценский с/с.
15.	ГИЛЕВСКИЙ Лев Леонидович	-"-	телефонист 785 СКП член ВЛКСМ	1925г.г. Минск, Серебрянка Юго-западная улица, дом 3	Бегомельским РВК	Умер 10.2.45г	-"	Бабушка Наталья Федоровна. г. Минск, Серебрянка, Юго- Западная улица, дом № 3.
16.	ШАНИНА Роза	старший сержант	снайпер 203 АЗОП канд ВКП/б/	1924г.г. Архангельск, Ленинградский 15.	Первомайским РВК	Умер 28.1.45г	Рхау Восточная Пруссия 3км с/з Ильмсдорф р-н г. Алленбург	Мать Анна Александровна г.Архангельск, Ленинградский, 15.
17.	ВАСИЛЬЧЕНКО Яков Семенович	крц	стрелок 747 ИПТАП ВЛКСМ	1926 г Красноярский край, Канский р-н, д. Круглово	Канским РВК Красноярского края	Умер 26.1.45	-"-	Мать Василиса Ефимовна. Красноярский край, Канский район, д.Круглово.
18.	ЖИГАЛКА Леонид Иванович	кр-ц	стрелок 612 СКП б/п	г. Минск. 2-я Линия дом 8.	Минским ГВК	Умер 25.1.45г	-"-	Жена Мария Лаврентьевна г. Минск, 2-я Линия, дом № 8.
19.	ТОЛЯРОНОК Павел Павлович	-"-	стрелок 449 СКП -"-	1921 г родных нигде нет		-"-	-"-	родных нет
20.	МЕТЛОВ Иван Михайлович	-"-	автоматчик 449 СКП канд ВКП/б/	1912г Ивановская обл, Маленковский р-н, д.Коровино	Маленковским РВК Ивановской области	Умер 26.1.45	-"-	Жена Анастасия Сергеевна Ивановская область, Маленковский р-н, д.Коровино
21.	ФАЙДЕЛЬ Гейниг Наймович	-"-	стрелок 785 СКП б/п	1903г г. Орша Красноармейская дом № 59.	Оршанским РВК	Умер 28.1.45г	Капиттен Восточная Пруссия р-н гор. Алленбург	Родных нет.

Начальник штаба 144 СВКОСД- полковник /ГОРЫЛЕВ/

Отп. 2 экз

Начальник 4-го отд.144 СВКОСД-майор а/с /УКРАИНЕЦ/

A casualty list, recording Roza's death (no. 16) on 28 January 1945, with her rank, unit and next-of-kin.

Roza's grave in the town of Znamensky in the Kaliningrad region.

January 14th, 1945

We began by crossing through Lithuania and Byelorussia. It's not as if our forces are unlikely to get far; our troops on the right captured Pilkallen yesterday, but today they have again been put under pressure. On the left they've gone a long way ahead. But artillery fire can still be heard. All morning we were listening to a rumbling cannonade. The Katyusha rockets announced the commencement of great events. Everyone went forward. No cart was left for our platoon, and I am going to bed, without anything. We can't carry stuff on our own backs, and yet we're mainly put in the position of having to go on foot. We've already run out of horses. Well, since we're not needed anywhere, nobody stands up for us. We went without supper and we didn't have breakfast. It's already 12 o'clock and we are still stuck here.

Yesterday I wrote to Vovik Emelyanov. I congratulated him on his marriage to my friend Sasha Yekimova. And I wrote that our trio (Sasha, Kalya and me) no longer exists. I am leaving it because our interests have diverged and we have nothing to talk about. I read it to Sasha. If I was living with the platoon, I would be friends with Valya Lazarenko, but I will soon be departing for the company. Valya and I made an agreement never to talk in front-line language and not to utter even a single non-literary word. Whoever breaks the agreement must hand over her sugar ration for a fortnight. The girls swear like troopers, but you can't do anything about it. My Sasha has let herself go.

To Nikolai I wrote: 'Wherever you are, my valiant falcon, I shall cherish my love for you.' I wrote that without him I don't have a guy because, if I met him now, he would not deceive me. I have become more resolute, but he so assured me back then that he loved me, and now he has forgotten, and I have even found out that he has a wife. I cursed him for his flattery and hypocrisy; I did not take offence at him not loving me.

In Valya's album there is a photo of Chelyabinsk. She is a pretty, young blonde. Her father almost drove her out of the house. She and I are alike, although I am less attractive, and she is more so, but she is less educated. She envies me for this. 'Yet true love is no wanderer; it gives a beauty that it does not find, and forges a chain no charms can break.' (*The Heart of Princess Osra*). On the second page she has a haircut like a boy, and is in trousers and a tunic, with a dog, like a boy – it's like it was at the front with the 159th Rifle Division in 1943. She's mad about horses, raves about them. 'To be in the cavalry, you must not only love it, but have the soul of a cavalryman,' – *Ivan Nikulin* [presumably the 1944 Mosfilm production depicting events in 1942]. After the war her ambition is to stay with the cavalry for the rest of her life.

We're standing there, the six of us, in Lithuanian garb, them and me. They were also a trio. Tanya was killed, Lyuda was wounded, and now she is on her own. Here they are, Lyuda and Tanya with Lyuda's father, in Valya's album. Here's a postcard – a beautiful woman with a horse; I gave it to her before she was wounded. Valya was also wounded; we came out of hospital together. I wrote: 'In memory of our days at the front, 24 October 44. Remember the march.'

We have just crossed into East Prussia. On what do we base our lives, we front-line girls? I have come to the conclusion that there is no such thing as truth; there are lies and hypocrisy. Remember how they looked at us … O girls, where can you turn to, where is the truth? It doesn't exist, even back then when I was advancing with the soldiers, and lies were spread about me, that I was behaving badly with the lads. And then 'You can die, but don't give a kiss without love.' Chernyshevsky. Here is her [presumably Valya Lazarenko's] composition. Waiting out arduous days. Days of separation from the family. Yours.

Remembering my yesterdays with longing,
I live on disenchanted with my fate.
I stood out above others of my kinfolk
As curious and covert in my thoughts.
Galloping and charging on horseback,
On my steed so playful and so wild.
Horse of mine, jet black in colour,
And silky and yellow of mane.
You were my sole joy so ardent.
How fine in the silence of night
To ride you, my faithful companion,
Midst splinters of mortars and shells
Darting through storms and infernos.
If you knew how I trust you, my friend,
My object of deepest affection.
If only you knew how to talk,
I would talk just to you and no other.
Now on my hospital bed
I simply cannot forget you
In a solitary, secretive way.
I am endlessly grieving for you,
So, do not be jealous, my beau.
You alone do I cherish,

But admit to you that at times
I do grieve a little for Genna,
Pine, though he loves me not.
I am sharing with you my sorrow
That he has another love.
He tells her: 'I love you only',
Though me he swore not to forget.
But that was of no long duration;
He is always prepared to forget.
To hell with it! I'll just forget all –
Genna, that first-time caress,
And come home to you, my comrade,
And you I'll caress ...

Evacuation Hospital 3049, Kaunas, January 12th. There is a frost outside. The cannon are unrelenting, coming closer and closer, on the right flank. We have been commanded to head off for Eydtkuhnen [now Chernyshevskoye]. We had a little to eat; we'd brought salami and bread with us. Our own artillery is warming up.

January 15th, 1945

After that we spent some time in the town of Eydtkuhnen, in the rear of the division. In the morning everyone's heading for the bath-house, but I put on new camouflage gear, kissed them all and now I am already in the rear by the general staff headquarters of the 144th Rifle Division; in an hour I'll be on the forward line. I have covered 20 km since morning. I wrote a nice letter to Nikolai B.: there's no reason to quarrel; maybe I'll be killed.

January 16th, 1945

The first night I spent at the general's. He gave me a fabulous reception, as did all the staff. Tomorrow I am heading out; I have come in search of the regiment. I met Kasimov, did not acknowledge him, and he did not recognise me. I met the lads on the self-propelled guns; we went on the attack in tanks. I was in a tank. One was put out of action, and some were seriously wounded. Major Tubanov lost his life. He was a friend of Sasha Yekimova, eight times decorated. Everybody was sorry.

In the evening I went to see Borovik and froze to the marrow. I made it to Borovik's dug-out. I wasn't as glad at seeing him as I was at getting into a warm dug-out. It was a heavy frost. My eyes smarted from being unused to the smoke. I can't breathe in these fumes. I slept like the dead.

Once again, the general has given me permission to stay in the forward zone. I went to the 216th Rifle Regiment and reported. I was taken but viewed suspiciously. They find it hard to believe I was permitted to go to the forward positions. The regimental commander is not allowing me through for the moment.

I have finally established that I am not capable of loving. What a tremor was awakened in my heart when I saw Nikolai B. for the first time. Today I have again found some shortcomings. There's a war on, but my heart grants no favours. Nikolai had no belt on his overcoat; the epaulette had been torn off, etc. I found this slovenly and my love was smothered by a feeling of revulsion. He's already become repulsive to me. He went into battle, and now he's been seriously wounded. A pity – that's war and all.

The news was passed on to the platoon that I was wounded, and they are not looking for me there. But I left for the regiment without the permission of Donyets [General Donyets, commander of the 144th Rifle Division].

Outside the wind is unbearable; the snowstorm is blowing not just snow, but dirt. The land is dry, and my camouflage suit is already giving me away; although it's dirty, it's too white. Today I have gone the whole day without a bite. My head aches from the tank fumes. I'm not registered for rations anywhere because I have no ration certificate, and for the moment I'm not listed anywhere. Back then I was half-hungry and today I'm really hungry. I'm not going to behave impertinently; I'll fast for a bit somehow. There's not long to go.

I'm being accepted as an outstanding sniper, and it seems that's the only reason I have been accepted. But everyone thinks that I came to this division only because I've got a guy here. The regimental commander even questioned me about it. I have resolved not to love anyone; I'm bound to be disappointed anyway. I'm here but I don't know a soul. I'm putting up with the dirt, cold and hunger. Everyone who knows me – from the tank crew to the general – is advising me to go back to my platoon, rather than endure a war like this: artillery fire, ear-shattering noise, the minute-by-minute destruction of my life.

I am very often under fire. On the forward line I was in the infantry, with the 785th Rifle Regiment under Commander Kasimov. While I was there, the girls were issued linen, fur coats and felt boots. How lovely and how warm.

They have attached me to a platoon of drivers for protection in the rear of the 157th Rifle Division. They ride in vehicle cabs, go see their dear ones. It's a warm, bright, well-fed existence. That's what I want too. But some sort of unknown force draws me on to the front. Oh passions, passions, oh blind strivings of the human heart. Onward, onward, it saith, and where beauty leads, there it follows [modified quote from *Sister Carrie*]. I submit to my heart. I like adventures, explosions, and it's particularly interesting to repel counter-attacks. Come what may, all in all, it's forward, a final advance with no turning back! But I want to eat; I've grown thinner over the last three days; I feel it myself.

It's now evening. So many casualties, but again we've gone forward a bit. We've moved only 10 km in five days. In three days the First Byelorussian Front advanced 60 km along a 120 km front. I sat down and had a think, and I will write again. Should I confine my will within the bounds of the law? It is the law which makes crawling snails of those who would soar like an eagle.

I am veiled in glory. The army newspaper *Unichtozhim Vraga* ['We shall destroy the enemy'] recently wrote: 'Having distinguished herself during an enemy counter-attack, Shanina is being awarded the medal "For Valour". She is an outstanding sniper of our sub-unit.' The Moscow magazine *Ogonyok* ['Flame'] featured my photo on its front page, saying 'She's wiped out 54 Germans, captured three, has two Orders of Glory' – that was earlier. I can imagine it: the whole country is reading it, all my friends, but who would know what I am feeling this minute?

Not long ago, Ilya Ehrenburg published a letter to me in a newspaper on behalf of our army, thanked Starostenko, the captain, the battalion commander, who was the first to set foot on German territory, and Yurgen too, and me, as an outstanding sniper. 'I thank her 57 times in a row; she has saved thousands of Soviet people.' But I thought to myself: surely that's not glory. Glory is splitting your own skull in the name of the motherland or smashing someone else's – that is glory (says *Bagration*), but this is what – just blather for folk in the rear, and in actual fact what have I done? No more than I was obliged to do as a Soviet citizen; I went to the defence of my homeland. Today I am agreeable to going into attack, even into hand-to-hand combat. I have no fear. I am sick of my life. I am glad to die in the name of my homeland. How good that I have that opportunity, or else I would have to die a foul death. Like many soldiers.

January 17th, 1945

I got up, didn't get a chance to have breakfast; the big chiefs arrived. I left for the battalion. I went on an assault with the infantry, in the front ranks. We moved too far ahead without informing the rear. As a result, we were hit by a Katyusha rocket and a heavy mortar. What a mess! It's the first time

I've experienced so much artillery fire. My first experience of machine-gun fire was on July 19th, beyond the Neman, with Solomatin. And now? Today seemed like a month to me. A shell landed right beside me and people were torn to pieces. I helped to bind up the wounded and carried on forward. Three of us burst into a house and fulfilled our set objective. But our division's route was changed; they veered to the left and our efforts turned out to be useless.

The 371st Rifle Division easily advanced behind us but it was impossible to go any further. The Fritzes were bombarding us with every kind of weapon. In the gully beyond the house, 100 metres away, there were enemy self-propelled artillery vehicles machine-gunning and shelling us. A Fritz glanced out of his hatch and I took him out from the house, and there were no more good targets for the rest of the day.

Frosts, hunger. I went back to my unit. Some of the lads were casting foul compliments my way. Obscene swearing everywhere. I'm so tired. I went to look for my own folk. I bumped into some friends, and went to search for the regiment. I came across the divisional command post and settled down for the night there. It was cold. I had a little to eat. In the house I grabbed some loot – a paper album into which I want to transcribe everything. How hard it all is! I can see that I'm of little use as a sniper. Maybe I'll have my moments, but death threatens. In our second battalion there are only six left out of 78. I pine for the girls; I live a much worse life than they do.

January 18th, 1945

I've been sitting here crying for three hours now. It's 12 o'clock at night. Who needs me? What use am I? I don't help with anything. Nobody's interested in my feelings. On the surface it seems I have an awful lot of sympathisers, but nobody can be relied on to offer help. I don't know what to do now. I often hear nasty things said. Why do I endure torments that are not good for anyone? Everyone bellows foul language, obscenities. I don't talk to anyone. Out of the blue someone asked: 'Is your name Shanina?' I didn't answer. It turned out it was Pavel Blokhin's best friend and I knew him well. I didn't recognise him now. What a pleasant meeting! He's the head of reconnaissance in the 785th Rifle Regiment. He said: 'I've been told what Shanina gets her medals for.' I confess it to Klava and I keep hearing bad reports. But I have taken a real liking to Junior Lieutenant Nikolai; he is attentive towards me.

Captain Tishin
Captain Vovka Stepanyenko
Captain Pavel Blokhin

Senior Sergeant Mishka Panarin 2
Senior Lieutenant Nikolai Solomatin
Senior Lieutenant Nikolai, gunner, 1st Battalion, 97th Regiment, 184th Rifle Division
Major Osmak 3–4
Lieutenant Vovka Ledison 3
Colonel Khorapov 3
Captain Nikolai Fyodorov
Senior Lieutenant Borovik

[There is no indication in the original as to why this list is included or what the significance of the numbers after some names may be.]

January 24th, 1945

I haven't written anything for ages. There was simply no time. I went to see the 785th Regiment reconnaissance team. They were great lads, gave me a good reception, but the chief of the regimental general staff began to pester me; I couldn't stand it and abused him, after which I spent another two days there and left. Any more would have been impossible; the harassment was intensifying.

Over these last two days there has been no time to take a deep breath. The fighting has been dreadful. The Germans manned and armed full trenches of infantry – they defended stoutly. Our troops drove over the trenches and took up a position in and around some farm buildings 150–200 metres from the trenches. The Fritzes opened fire when our lads were going past. It was a real massacre. Our troops were being carried on self-propelled guns and tried to take them to the objective. Just one or two made it; the rest were destroyed by fire. I rode on a self-propelled gun, but did not manage to fire from it. You couldn't stick your head out through the hatch; people were being wounded and killed. I went along the gully, crawled out and fired at the Fritzes fleeing from the trench.

By evening we had driven them all out, occupied the objective and come across an anti-tank ditch. I am pushing on, but the infantry are stationary, afraid to go further. Two scouts from a penal battalion were moving forward. I went with them and as a result we three were the first to occupy the next farm, and all those behind us started advancing too and began to snap at the heels of the fleeing Fritzes. I fired along with all the others. It turned out that these penal troops were our neighbours on the left, from the 63rd Rifle Division.

On seeing me, the commanders of the 63rd shouted to the men: 'Follow the example of that girl. Learn from her.' They tried to keep me with them, but I went to look for my own unit. I was running and shouting to the soldiers: 'What division are you?' I heard behind me soldiers crying: 'Halt!' And to my left two Fritzes got up from behind some bushes and walked towards me with their hands up, just a few metres away from me.

I met the divisional scouts. They took me in and said: 'You can stay with us.' And we headed off westward. We picked up 14 lost Fritzes. We're already marching ahead. The Fritzes are fleeing helter-skelter. Suddenly the order came: to go back and to the right. We were on motor transport, proceeding in columns, heading for the town of Schlossberg [Dobrovolsk]. We passed through the town and moved on further. Here the Germans ditched everything – cows, the whole lot – and fled to the forest. The village is being bombarded. *Fraus* are coming out. There are a lot of Lithuanians. But you should see the machinery we've got! Lord, the whole army is on the move; there are complaints people aren't observing the traffic rules.

There's a big iron bridge over a stream, and a beautiful highway running above the level of the fields. Near the bridge some trees have been hacked down, but they didn't have time to create an obstruction. The houses are lavish, built of stone; everywhere the décor is splendid: piano, mirror on the wall, blinds of silk, plush and tulle, fancy armchairs, and all the furniture as well. The scouts haven't time for me; they're occupied by their work, and there's nowhere to sleep. They've abandoned us.

I have been with the division. Vadim, the son of the colonel and chief of general staff, is a lieutenant. He doesn't do anything; he's a mummy's boy and kind of nasty.

Another night march. It's dark now. It'll soon be first light. I'm sitting by the campfire, writing. How unfortunate that I have nobody placed over me. It's good that no one will order me about, but not so good that there is nobody to suggest what to do. My heart can find no satisfaction. Nobody needs me.

End of Roza's Diary.

AND THIS WAS HOW SHE LOST HER LIFE

Marat Yegorovich Shanin writes: Private Nikolai Vasilyevich Lentsov wrote this in a letter to me thirty years ago:

'We ran over in response to a heart-rending woman's scream. Roza was lying on the ground, her sniper's rifle beside her. She was shouting: "Lads! Finish me off quickly!" With both hands she was holding in her innards, that were spilling out as a result of a shell splinter which had torn her belly open.'

He bound her up, assisted by Private Leonid Vasilyevich Dubov. Together they carried Roza away from the scene of the fighting.

From a letter by nurse Yekaterina Petrovna Radkina:

'I shall always remember Roza. It's impossible to forget her achievements. We were not close acquaintances. For us nurses she was a celebrated heroine. Her life was a model of valour and courage.

The first time Roza came to us on account of a wound in the shoulder, but not for long.

Sometimes I saw her from a distance in the political section (I was secretary of the Young Communist League branch for the 205th Medical Team, in the 144th Rifle Division).

Our last meeting, I will remember for the rest of my days. There was very heavy fighting. The forces were advancing quickly. In order to keep up with the division, the medical team was strung out in stages. I had to stay with the seriously wounded, to transfer them on to hospitals. I had handed over a group of wounded and was catching up with the medical personnel in Reichau [Telmanovka]. But they had already left in pursuit of the troops and again I was left with a very large number of seriously wounded, among whom I saw Roza. Roza was very badly wounded in the gut. She was in an extremely grave condition. But there were no groans or tears. She was very thirsty, but drinking was prohibited. She begged me: "Katya! Just give me a drop of good old cold water. I'll just rinse my mouth out!"

The wounded who were in action with Roza were full of admiration for the way she had stood in for their dead commander and led them into battle.

Roza realised the gravity of her condition. She knew she would not survive. She regretted that she had managed to accomplish so little. She remembered her relatives and called for her mother. I was with her to the last. I shall remember Roza as lively, steadfast and brave. She is a true hero.'

'Yekaterina Petrovna told me in Moscow,' Marat Yegorovich recalls, 'about the spring of delicious cold water in Bogdanovskoye, which Roza mentioned on the last day of her life'.

On February 26th, 1976, Marat Shanin received a letter from Valya Lazarenko (married name, Rogova). It was with her that Roza had made their pact not to use a single non-literary word, with any violation resulting in the offender forfeiting her sugar ration for a while. Valya Lazarenko was wounded in the leg in battle near Pilkallen. She writes:

> 'You and your children ought to feel proud that your Roza was an outstanding comrade, a bold and fair-minded soldier, whose example we all followed. I was not a close friend of hers at first, but became one after my return from hospital. But by that time Roza had lost her sniping partner. She offered me her friendship. And we agreed that, as soon as I was able to go out "stalking" (I had not completely recovered in hospital and my leg still bothered me), we would form a sniper partnership. But this was destined never to be. Roza went off to a company on her own. I was not with her, but then we were told how bravely she had behaved. Her commander was killed in action, Roza stood in for him, but was wounded and died in hospital. This was in January 1945.'

Roza Shanina was buried on the high bank of the River Lava [Alle in German, Lyna in Polish], not far from the town of Gvardyesk, in the village of Bolshaya Polyana, Znamensky District. This bank of the Lava would have been the last thing Roza saw. What was she thinking? What life pictures would have sped so rapidly through her consciousness? Did the River Ustya of her childhood flow past her dimming vision? Did the cranes from her childhood with their mournful reedy call bear her away beyond the horizon and melt away in the deep blue sky. It is an open place. Along the bank of the Lava runs a strip of forest. The ruins of a brick works are barely noticeable. The road leading to Roza's grave is lined with trees, and the road itself is of red clay. Indeed, it is known as Red Road. Up to 1965 the residents of Telmanovka believed that it was a German woman who was buried under the spreading pear tree on the bank of the Lava.

Lidiya Vladimirovna Melnitskaia was the first to find Roza Shanina's grave in 1965, while on an assignment from the regional newspaper *Severny Komsomolets* ['The Northern Young Communist'].

Among those who lost their lives in combat for the homeland were Aleksandra Koreneva, Aleksandra Yekimova, Anna Nesterova, Lyubov Tonailova and other women snipers who fought on the Third Byelorussian Front.

In recent times Roza Shanina's tomb was transferred to the town of Znamensky, Kaliningrad Region, and is located beside a common grave where 400 defenders of the homeland are interred.

V. P. Mamonov, N. N. Poroshina
(authors of *She Left us a Legacy of Songs and Dewdrops*)

BIBLIOGRAPHY

Nevill Armstrong: *Fieldcraft, Sniping and Intelligence*. Gale & Polden, 1942, third edition.

Charles Bean: *Official History of Australia in the War of 1914–1918: Volume I - The Story of Anzac: The First Phase*. Australian War Memorial, 1921.

—— *Official History of Australia in the War of 1914–1918: Volume II - The Story of Anzac: From 4 May, 1915, to the Evacuation*. Australian War Memorial, 1924.

Alla Begunova: *Angely Smerti: Zhenshchni-snaĭpery, 1941–1945* ('Angels of Death: Women Snipers, 1941–1945'). Veche, 2014.

Frederick M. Crum: *Memoirs of a Rifleman Scout*. Frontline Books, 2014.

Denis Edwards: *The Devil's Own Luck*. Leo Cooper, 1999.

John George: *Shots Fired in Anger: A Rifleman's View of the Battle of Guadalcanal*, Canton Street Press, 2013 reprint of 1947 edition.

Adrian Gilbert: *Sniper: One-on-One*. Sidgwick & Jackson, 1994.

—— *Stalk and Kill: The Sniper Experience*. Sidgwick & Jackson, 1997.

—— with Tom C. McKenney, Dan Mills, Roger Moorhouse, Tim Newark, Martin Pegler, Charles W. Sasser, Mark Spicer, Leroy Thompson and John B. Tonkin: *The Sniper Anthology: Snipers of the Second World War* (Introduction by John L. Plaster). Pen & Sword, 2012.

John Hamilton: *Gallipoli Sniper: The Remarkable Life of Billy Sing*. Frontline Books, 2015.

Michael E. Haskew: *The Sniper at War: From the American Revolutionary War to the Present Day*. Thomas Dunne, 2005.

Hesketh Vernon Hesketh-Prichard: *Sniping in France, With Notes on the Scientific Training of Scouts, Observers, and Snipers*. Hutchinson & Co., 1920.

Roland Kaltenegger: *Eastern Front Sniper: The Life of Matthäus Hetzenauer*. Greenhill Books, 2017.

Richard D. Law: *Sniper Variations of the German K98k Rifle*. Collector Grade Publications, 1996.

Herbert W. McBride: *A Rifleman Went to War: How One Man Revolutionised Sniping in World War One*. Endeavour Press, 2016.

Evgeniy Nikolaev: *Red Army Sniper: A Memoir of the Eastern Front in World War II*. Greenhill Books, 2017.

—— and Dina Nikolaeva: *Sovetskie Snayperu v Boyo 1941–1945*. Moscow, 2009.

Lyudmila Pavlichenko: *Ya snaïper* ('Me: The Sniper'). Moscow, 2014. (Published in English in 2018 by Greenhill Books, as *Lady Death: The Memoirs of Stalin's Sniper*).

Martin Pegler: *Out of Nowhere: A History of the Military Sniper*. Osprey Publishing, 2004.

—— *Sniping in the Great War*. Pen & Sword, 2008.

—— *Sniping Rifles on the Eastern Front 1939–45*. Osprey Publishing, 2019.

Eric Penberthy: 'British Snipers (i). An Account of the Training and Organisation of Snipers in the British Armies in France', *The English Review*, September 1920.

Joseph Pilyushin: *Red Sniper on the Eastern Front*. Pen & Sword, 2010.

John L. Plaster: *The History of Sniping and Sharpshooting*. Paladin Press, 2008.

Tapio A.M. Saarelainen: *The White Sniper Simo Häyhä*. Casemate Publishing, 2016.

Charles W. Sasser and Craig Roberts: *One Shot - One Kill*. Pocket Star Books, 1990.

Peter R. Senich: *The German Sniper*. Paladin Press, 1996.

Clifford Shore: *With British Snipers to the Reich*. Frontline Books, 2012.

Ian Skennerton: *The British Sniper*. Skennerton Publishing, 1983.

—— *The Lee-Enfield Story: The Lee-Metford, Lee-Enfield, S.M.L.E. and No. 4 Series Rifles and Carbines, 1880 to the Present*. Greenhill Books, 1993.

Frederick Sleath: *Sniper Jackson*. Herbert Jenkins Ltd, 1919; reprinted Frontline Books, 2014.

Mark Spicer: *An Illustrated Manual of Sniper Skills*. Pen & Sword, 2016.

Bruno Sutkus: *Sniper Ace: From the Eastern Front to Siberia* (Introduction by David L. Robbins). Frontline Books, 2009.

Andrey Ulanov: *Stalkers of the Enemy: Red Army Snipers*. To be published by Greenhill Books in 2022.

Lyuba Vinogradova: *Avenging Angels: Soviet Woman Snipers on the Eastern Front (1941–45)* (Introduction by Anna Reid). Maclehose Press, 2017.

Albrecht Wacker: *Sniper on the Eastern Front: The Memoirs of Sepp Allerberger Knight's Cross*. Pen & Sword, 2016.

Glenn Wahlert and Russell Linwood: *One Shot Kills: A History of Australian Army Sniping*. Army History Unit, 2014.

John Walter: *Guns of the Reich: The Small Arms of Hitler's Armed Forces, 1933–1945*. History Press, 2016.

—— *Snipers at War: An Equipment and Operations History*. Greenhill Books, 2017

—— *The Sniper Encyclopaedia: An A–Z Guide to World Sniping*. Greenhill Books, 2019.

Vasiliy Zaytsev: *Notes of a Russian Sniper: Vassili Zaitsev and the Battle for Stalingrad*. Frontline Books, 2015.

Yuliya Zhukova: *Girl with a Sniper Rifle: An Eastern Front Memoir*. Greenhill Books, 2020.